Port State Control Practice

Volkan Aydogdu 편저
박영수
김종성
김종수
임태우
이윤석
윤귀호
이형기
배병덕

도서출판 상학당

머 리 말

항만국통제(Port State Control, 이하 PSC)란 선주, 선급 및 기국(Flag State)이 국제 협약에서 요구하는 사항의 이행 여부를 선박이 입항하는 항구의 항만 당국이 확인하고 점검하며, 또한 협약의 이행에 대한 모든 권한이 기국에 있음에도, 당해 항만 당국은 발견된 결함사항에 대하여 출항하기 전까지 시정을 요구할 수 있는 권한을 가진 것을 말한다.

PSC는 협약의 이행 여부를 확인하는 데 있어서 기국과 항만 당국이 상호 보완관계에서 항만 당국(Port Authorities)은 자국의 안전을 위하여 협약의 준수 여부를 점검하는 권한을 가지고 있으며, 기국은 자국선에 대하여 협약을 준수할 의무가 있다.

최근 환경의 중요성이 전 세계적으로 확산하면서 PSC의 중요성도 점점 확대되고 있으며, 이는 항만 당국으로 하여금 PSC의 점검 강화를 유도하고 있다.

PSC 검사를 통해서 출항 정지율이 높은 선사, 기국 및 선급 등을 분류하여 인터넷에 명단을 공개, PSC 우선 점검실시와 입항거부 등 직접적인 불이익을 주고 있다.

이 저서 PSC 실무는 현직에서 종사하는 PSC 검사관과 교육기관의 학생들을 대상으로 PSC 검사중 반드시 필요한 여러 가지 사항들을 국제 언어인 영어로 집필하여 실무에 많은 도움이 되도록 구성하였다.

부족한 저서이지만 아무쪼록 이 책이 관련 종사자들에게 많은 도움이 되길 바라며, 앞으로도 지속적인 개정 및 보완을 할 것임을 다짐하는 바이다.

끝으로, 이 교재의 편집과 출판에 도움을 주신 도서출판 상학당 관계자분들에게도 감사의 말씀을 전하는 바이다.

2009년 2월

편저자 씀

차 • 례

Definitions Used In Connection With Port State Control

Clear grounds: Evidence that the ship, its crew or its safety management system does not comply with the requirements of the relevant conventions. Such evidence needs not necessarily be a deficiency, but may be an incident, and accident or and indication of substantial non-compliance/detainable deficiencies.

Deficiency: Non-compliance, discrepancy or deviation from the requirements of the relevant instruments/conventions.

Detainable Deficiency: A deficiency that presents an immediate threat to the ship, its personnel or the environment, which renders the ship unsafe to proceed to sea.

Detention: Intervention action taken by the port State in case of detainable deficiencies or substantial non-compliance to ensure that the ship does not sail until detainable deficiencies have been rectified.

Expanded inspection: An inspection conducted according to non-mandatory guidelines only once during 12 months period for certain types of ships (tankers, bulkers and passenger ships) and certain categories of age and size.

Inspection: A visit on board a ship to check both the validity of relevant certificates and other documents, and the overall condition of the ship, its equipment, and its crew.

More detailed inspection: An inspection conducted when there are clear grounds for believing that the condition of the ship, its equipment, or its crew does not comply with the requirements of the relevant conventions. The inspection may focus on one area or be across various areas.

Port State Control Officer (PSCO): A person duly authorised by a Port State authority to carry out port State control inspections, and responsible exclusively to that authority

Recognised organisation: An organisation which meets the relevant conditions set forth by resolution A.739(18), and has been delegated by the flag State Administration to provide the necessary statutory services and certification to the ships entitled to fly its flag.

Serious Deficiency: A deficiency that may present an immediate threat to the ship, its personnel or the environment, and that could be qualified as a detainable deficiency.

Stoppage of an operation: Formal prohibition against a ship to continue an operation due to identified deficiency that, singly or together, render the continuation of such operation hazardous.

Substandard ship: A ship whose hull, machinery, equipment, or operational safety is substantially below the standards required by the relevant convention or whose crew is not in conformance with the safe manning document.

Valid certificate: A certificate that has been issued directly by a Flag Administration or on its behalf by a recognised organisation and contains accurate and effective dates; meets the provisions of the relevant convention; and with which the particulars of the ship, its crew and its equipment correspond.

Action Codes: The given time frame for rectification of each deficiency is commonly given in a coded form in the inspection report, called "action code". Following codes are mainly used:

- 30 = Grounds for detention
- 17 = Master instructed to rectify deficiency before departure
- 16 = to be rectified within 14 days
- 15 = to be rectified at next port of call
- 19 = rectify major non-conformity before departure
- 18 = rectify non-conformity within 3 months
- 10 = deficiency rectified
- 40 = next port informed
- 50 = Flagstate/consul informed
- 70 = Classification society informed

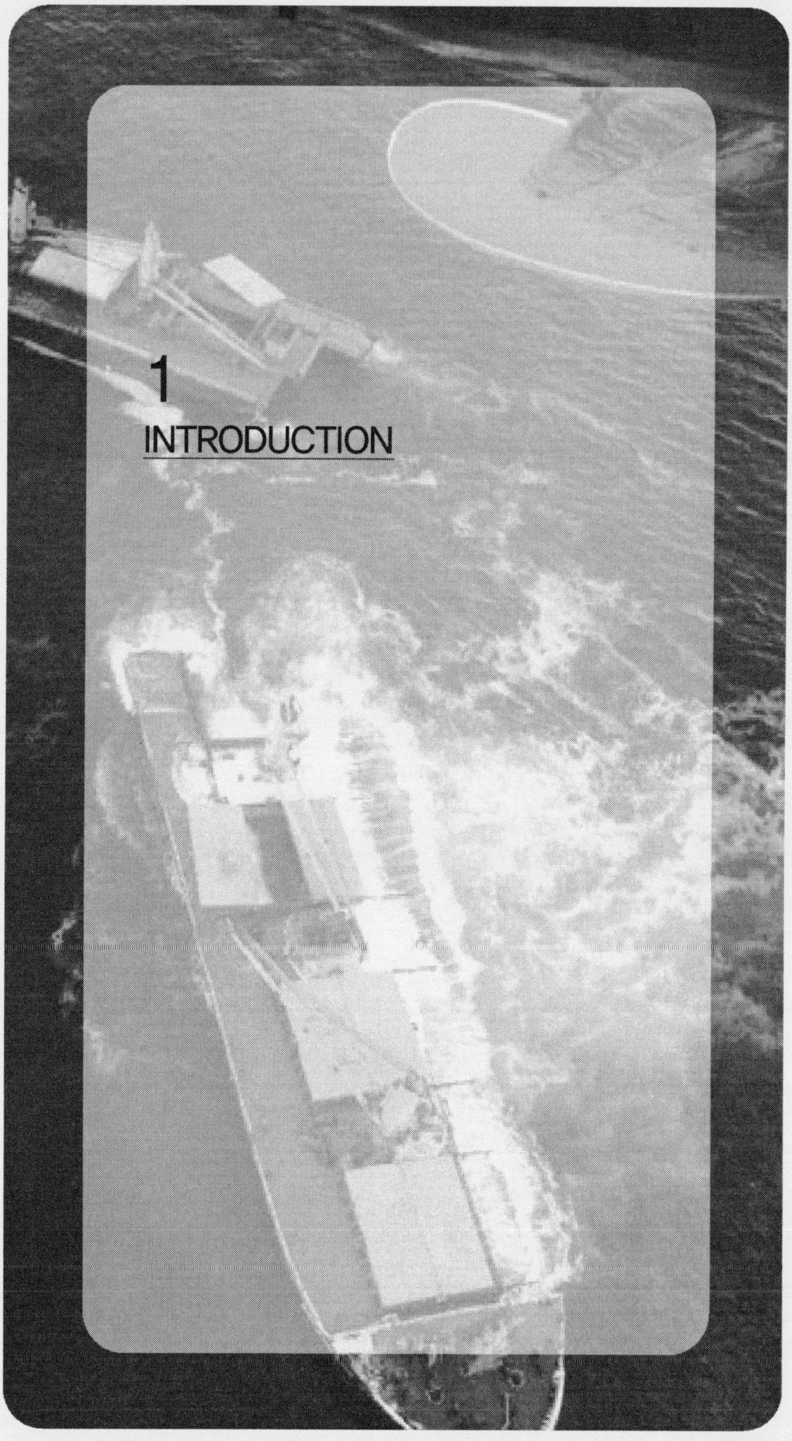

1
INTRODUCTION

1 - INTRODUCTION

Port State Control - what is it and why is it necessary?

The United Nations Convention of the Law of the Sea (UNCLOS) provides every nation with many rights with regards to vessel registration and freedom of passage both over the high seas and through coastal waters of any other nation. With these rights comes certain responsibilities and these responsibilities remain with the nations who enjoy the rights. The responsibilities are set out in a variety of International Conventions, which have been developed over many years, most recently under the auspices of the International Maritime Organization (IMO). The most commonly accepted Conventions are the International Convention for the Safety of Life at Sea (SOLAS), the International Convention for the Prevention of Pollution from Ships (MARPOL), the International Convention on Load Lines, the International Convention on the Standards of Training, Certification and Watchkeeping for Seafarers (STCW) as well as numerous technical Codes and Resolutions associated with these Conventions.

The Administration offering vessel registration is referred to as the flag State, and is obliged to ensure that every vessel registered in their country complies with all the applicable Conventions, Regulations and Codes depending on the vessel type and trade routes. This task is a major undertaking for flag States with limited resources. It can lead to some flag States not carrying out all the necessary checks to ensure that all their vessels continue to comply with the relevant requirements. In practice, this has led to the majority of flag States delegating the responsibility of the checks to "recognised organizations" which are most commonly classification societies. Such societies have developed large networks of worldwide resources to enable them to carry out these delegated tasks but there can be a conflict since the societies have become commercial service providers paid for by the vessel operators.

The vessel operator is also critical in maintaining the required standards. Where an operator accepts this responsibility, and seeks to provide the necessary management and resources to enable a ship to comply with the international Conventions, the role of the flag State becomes secondary. A good operator in conjunction with a leading classification society can comply with the necessary international Conventions with minimal involvement by the flag State.

Operators can, however register their vessels with flag States that fail to meet their regulatory obligations and who provide minimal oversight. When this occurs another country can find ships arriving at its ports that are unsafe and threaten the marine environment. That country as "port State" has the right under the same international Conventions described above to carry out its own checks and to intervene to ensure that ship does not continue to pose a threat to safety or the environment. This is port State control, and it has assumed prominence in the shipping industry, driven by the consistent failure of the other responsible parties to fully meet their obligations.

As a summarize; **Port State Control (PSC)** is the inspection of foreign ships in national ports to verify that the condition of the ship and its equipment comply with the requirements of international regulations and that the ship is manned and operated in compliance with these rules.

Most important technical conventions of IMO contain provisions for ships to be inspected when they visit foreign ports to ensure that they meet IMO requirements

These inspections were originally intended to be a back up to flag State implementation, but experience has shown that they can be extremely effective, especially if organized on a regional basis. A ship going to work on international trade will normally visit other countries, in the region before embarking on a vessel, it is to everybody's advantage if inspections can be closely co-ordinated. This ensures that as many ships as possible are inspected but at the same time prevents ships being delayed by unnecessary inspections. The primary responsibility for ships' standards rests with the flag State - but port State control provides a "safety net" to catch substandard ships.

IMO has encouraged the establishment of regional port State control organizations and agreements on port State control -of Understanding or MOUs - have been signed covering all of the world's oceans: Europe and the north Atlantic (Paris MOU); Asia and the Pacific (Tokyo MOU); Latin America (Acuerdo de Viña del Mar); Caribbean (Caribbean MOU); West and Central Africa (Abuja MOU); the Black Sea region (Black Sea MOU); the Mediterranean (Mediterranean MOU); the Indian Ocean (Indian Ocean MOU); and the Arab States of the Gulf (GCC MoU (Riyadh MoU)

In accordance with the international regulations stipulated by International Conventions in the maritime field the main responsibility for ship safe condition is addressed to the flag State – the State under which flag the ship is registered.

Port State control (PSC) comes into the scene when shipowners, classification societies and flag State administrations have failed to comply with the requirements of the international maritime conventions. Although it is well understood that the ultimate responsibility for implementing conventions is left to the flag States, port States are entitled to control foreign ships visiting their own ports to ensure that any deficiencies found are rectified before they are allowed to sail.

Summarized reasons failed to comply with conventions

- fast growing of shipping industry in the world
- quick increasing number of vessel and tonnage
- spreading easy flag implementation
- vessels are working away than flag state

- some flag state could not impel ship owner to comply with conventions because of commercial care
- politic and social press and effect
- easy classification occurred and spread
- commercial cares
- vessels get older and new ship construction is expensive

Ratio of increasing easy flag implementation as GT

- **1938** % 1
- **1953** % 6.2
- **1973** % 22.8
- **1983** % 25.3
- **1993** % 40.1
- **1998** % 47.8

Finally as a result of deterioration in marine commercial filo :

- flag states lost their control on vessel
- the vessel which have to be controlled frequently by owner, flag state, charterer, shipper, classification, P&I...start working relatively without control and without comply international regulations
- these vessel which without control threat environment, life and cargo
- number of accident happened
- As a result insurance payment increased astronomically

2
MOUs around the world

2 - MOUs around the world

2.1 First Memorandum of Understanding

In March 1978, a massive oil spill occurred off the coast of Brittany (France), as a result of the grounding of the supertanker 'Amoco Cadiz. These incident caused a strong political and public outcry in Europe for far more stringent regulations with regard to the safety of shipping. This pressure resulted in a more comprehensive memorandum which covered:

- safety of life at sea
- prevention of pollution by ships, and
- living and working conditions on board ships.

Subsequently, a new, effective instrument known as the Paris Memorandum of Understanding on Port State Control was adopted in January 1982 and was, initially, signed by fourteen European countries. It entered into operation on 1 July 1982. Since that date, the Paris Memorandum has been amended several times to accommodate new safety and marine environment requirements restrain from the International Maritime Organization (IMO)

Legal basis from conventions:

Solas 74 CI R 19 (Control)

"Every ship when in a port of another Contracting Government is subject to control by officers duly authorized by such Government. The officer carrying out the control shall take steps to ensure that the ship shall not sail until it can proceed without danger to the ship or persons on board.

- SOLAS CI R19
- SOLAS CIX R6.2
- SOLAS CXII R4
- ILCC 66 ARTICLE 21
- MARPOL ARTICLES 5&6
- MARPOL AN I R8A
- MARPOL AN II R15
- MARPOL AN III R8
- MARPOL AN V R8
- STCW 78 ARTICLE
- LL 69 ARTICLE 12

2.2 Structure of MOU

The PSC Committee (PSCC)is the highest and decision making organ. The Maritime Advisory Board (MAB) and the Technical Evaluation Group (TEG) are advisory to the PSCC on respectively policy and technical matters. There are also permanent secretaries for MOU's and a central database and information systems such as SIRENAC established for most of the MOU. For example, SIRENAC is a private database and information system, with a mailbox to exchange also messages between the different ports. There is no access for outsiders, but the inspection and detention records for a ship can be found on the Paris MOU website and are also passed on to the EQUASIS database. Same as Paris MOU, other MOUs also have their own database and information systems or their systems are under construction.

Being the executive body of a MOU, PSCC is composed of the representatives of the participating maritime authorities and relevant organizations, such as representatives of IMO and ILO participate as observers in the meetings of the

PSCC, as well as representatives of several cooperating maritime authorities and other regional agreements on PSC.

Basic Principles

The prime responsibility for compliance with the requirements laid down in the international maritime conventions lies with the shipowner/operator. The responsibility for ensuring such compliance remains with the flag state.

Each maritime authority gives effect to the provisions of the relevant MOUs.

Each authority has to ensure that foreign merchant ships visiting its ports comply with the standards laid down in the relevant conventions and all amendments thereto in force. In this context, a participating maritime authority regards a ship flying the flag of another member state as a foreign ship too.

The MOUs provide for a total number of inspections, expressed in terms of a percentage, that each of the states party to the relevant MOU shall conduct. Under the Paris MOU the member states have agreed to inspect 25%of the estimated number of individual foreign merchant ships which enter their ports during a 12 month period. This percentage is different in other MOUs.

IMO and ILO conventions provide the basis for inspections under the MOUs.

All possible efforts are made to avoid unduly detaining or delaying a ship.

In principle, there will be no discrimination as to flag.

Inspections are generally unannounced.

In general ships will not be inspected within six months of a previous inspection in a MOU port, unless there are "clear grounds" for inspection

PSC is carried out by a Port State Control Officer (PSCO). The PSCO's powers derive solely from the sovereign state which employs him and is subject to the national laws of the jurisdiction in which he is operating. The PSCO should be an experienced person qualified as a flag state surveyor and able to communicate with the master and key crew members in English.

2.2.1 MOUs around the world

In order to carry out PSC effectively, “"the recommendation concerning regional cooperation in the control of ships and discharges”" was adopted by the IMO as a resolution. In July 1982, European countries signed the Paris Memorandum of Understanding on Port State Control (Paris MOU), and today many of the countries have signed and accepted MOUs. Currently, nine MOUs exist in the world and their status in implementing PSC is described below.

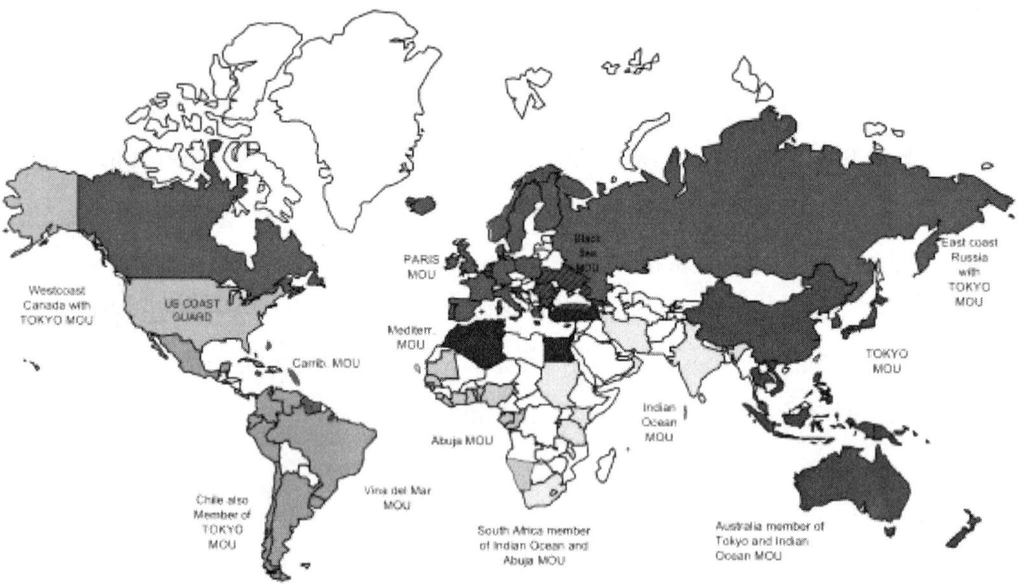

Distribution of PSC MOUs worldwide

(1) European region (Paris MOU)

Established: 1 July 1982

Members: Belgium, Canada, Croatia, Cyprus, Denmark, Estonia, Finland, France, Germany, Greece, Iceland, Ireland, Italy, Latvia, Lithuania, Malta, Netherlands, Norway, Poland, Portugal, Russian Federation, Slovenia, Spain, Sweden and United Kingdom

Note:

Under the agreement each country undertakes to inspect 25% of individual foreign flagged ships visiting their ports, to pool inspection information and harmonize procedures. The co-ordinated effort results in inspection coverage of 90% to 100% of individual ships visiting the region.

The Paris MOU Port State Control Committee (PSCC) held its 34th meeting in Bruges, Belgium, from 8 to 11 May 2001. It was agreed that a dialogue should be opened with the Tokyo MOU and United States Coast Guard (USCG) with a view to establishing harmonized criteria to evaluate the performance of classification societies. Recognizing the need for the introduction of a reward system to encourage quality operators, the Committee instructed a Task Force to develop such a system.

The Committee also instructed Task Forces to undertake a detailed analysis of deficiencies related to the human element and consider further harmonization of detention criteria. The Committee agreed on a common policy to respond to appeal/complaints from flag States, classification societies and operators and to aim to make this response within 10 working days.

The Paris MOU Advisory Board decided on 28 January 2002 to take account of the recommendations of the IMO Sub-Committee on STCW. If a seafarer's documentation did not comply with STCW95 then a Letter of Warning was issued and details of the ship were published on the Paris MOU internet site. Letters of Warning were issued until 31 July 2002. From the beginning of 2002 August a ship to which a Letter of Warning has been issued is to be subject to priority inspection and may be detained if the documentation of the crew does not comply with the requirements of STCW95.

The Paris MOU Committee 35th meeting was held in Halifax, Canada from 4 to 10 May 2002. In a major review of the Memorandum they agreed on banning procedures which could result in 'two strikes and out' for many sub-standard ships. The banning provisions mean that oil, chemical and gas tankers, bulk carriers and passenger ships flying a flag featured on the MOU's 'black list' and with a poor detention record will be refused access to MOU ports. Ships from flags in the 'high' and 'very high' risk categories will be banned if they are detained more than once in a three year period.

For ships from 'medium' to 'medium to high' risk flags the ban will take effect after the third detention in two years. The black list is available in the Annual Report on the Paris MOU website.

Also added to Memorandum is the International Labour Organisation's Protocol to ILO 147 which covers the checking of new requirements for seafarers hours of work and rest. These provisions entered into force on 22 July 2003. The meeting confirmed its intention to carry out a 3-month concentrated inspection campaign (CIC) on the International Safety Management (ISM) Code from 1 July 2002 when all ships required to have safety management systems in place.

(2) Asia-Pacific region (Tokyo MOU)

Established : 2 December 1993

Members : Australia, Brunei Darussalam*, Canada, Chile, China, Fiji, Hong Kong, Indonesia, Japan, Korea, Malaysia, New Zealand, Papua New Guinea, Philippines, Russian Federation, Singapore, Solomon Islands, Thailand, Vanuatu and Vietnam

Note:

The project for exchanging information between the database of the Tokyo MOU, APCIS and the database of the Paris MOU, SIRENAC has been completed and started operation in April 2001. Through this inter-regional data exchange, PSC officers in both regions can access the databases of each other to make searches and view inspection details of ships by way of the Internet.

The tenth meeting of the Tokyo MOU Port State Control Committee (PSCC) took place in Tokyo, Japan from 15 to 18 October 2001. The PSCC agreed upon a number of amendments to the text of the MOU and the PSC Manual. In an effort to harmonize regional PSC procedures, one inter-sessional group was tasked to incorporate some of the contents of the Paris MOU Manual into the Tokyo MOU Manual and another inter-sessional group on target factors was re-established to liase with the Paris MOU to ensure that target factors in both regions were as close as possible, all the while ensuring that they met the particular requirements of the regions. The PSCC noted that preparatory work for the conduct of a concentrated inspection campaign (CIC) on ISM Code implementation scheduled for July-September 2002 had been completed. During the Committee meeting, an open forum was organized to exchange views between the PSCC members and industry representatives on matters of mutual interest. The International Association of Classification Societies (IACS), Japan Shipowners' Association and ClassNK attended this session.

The eleventh meeting of the Tokyo MOU PSCC took place in Manila, the Philippines from 10 to 13 June 2002. The PSCC welcomed Chile as the eighteenth member of the Tokyo MOU. The PSCC confirmed the arrangements for the CIC on the ISM Code compliance. The PSCC approved the Tokyo MOU ship targeting system. The element of the targeting system is basically similar to the targeting factors used by the Paris MOU. The PSCC decided to start to implement the targeting system on a trial basis from the beginning of 2003 and .....

(3) Latin-American region (Vina del Mar or Latin-America Agreement)

Established : 5 November 1992

Members : Argentina, Bolivia, Brazil, Chile, Colombia, Cuba, Ecuador, Honduras, Mexico, Panama, Peru, Uruguay and Venezuela

Note:

The Vina del Mar Agreement held its eighth PSCC meeting in Mexico, from 3 to 5 September 2001. The agenda included the control of fishing craft and passenger ships, procedures for the notification of detentions to the captain of the detained ships, safe cargo stowage and securing procedures, training of Port State Control Officers (PSCO), status of relevant conventions under the Agreement and relevant decisions of other regional agreements and international organizations.

(4) Caribbean region (**Caribbean MOU**)

Established: 9 February 1996

Members: Anguilla, Antigua and Barbuda, Aruba, Bahamas, Barbados, Belize, British Virgin Islands, Cayman Islands, Commonwealth of Dominica, Cuba, Grenada, Guyama, Haiti, Jamaica, Montserrat, Netherlands Antilles (Curacao, St.Maarten), Suriname, St. Kitts Nevis, St. Lucia, St. Vincent & the Grenadines, Turks & Caicos Island and Trinidad & Tobago

(5) Mediterranean region (**Mediterranean MOU**)

Established: 11 July 1997

Members: Algeria, Cyprus, Egypt, Israel, Jordan, Lebanon, Malta, Morocco, Tunisia, Turkey and Palestinian Authorities

Note:

The fourth PSCC meeting of the Mediterranean MOU took place in Valletta, Malta, from 14 to 18 July 2001. The Committee accepted and thanked the European Commission for its offer to provide the Secretariat with an advanced training programme for PSCOs on CD ROM for use by Member States of the Mediterranean MOU. The Committee agreed that charges for verification visits should only be levied if they were carried out outside normal working hours at the request of the shipowners/operators/agent; that a Sub-Committee on Detention Disputes (SCDC) should be set up within the Mediterranean MOU to consider appeals by flag States; and that ships which had been subject to an expanded inspection by PSC authorities of the Paris MOU, Tokyo MOU and United States Coast Guard within six months prior to their calling at ports of the Mediterranean MOU would not have to be reinspected by the Member States of the Mediterranean MOU, unless clear grounds were established which warranted or justified their further inspection.

(6) Indian Ocean region (Indian Ocean MOU)

Established: 5 June 1998

Members: Australia, Bangladesh, Djibouti, Eritrea, Ethiopia, India, Iran, Kenya, Maldives, Mauritius, Mozambique, Myanmar, Oman, Seychelles, South Africa, Sri Lanka, Sudan, Tanzania and Yemen

Note:

The fourth session of the Indian Ocean MOU (IOMOU) PSCC was held in Colombo, Sri Lanka, from 3 to 6 September 2001. The Committee discussed Certificates of Competency issued by an Authority not included in the list of parties that have complied with all the requirements of STCW Convention and decided that if the flag State, on whose ships these persons are employed, accept these certificates, the same should be accepted for PSC purposes. The Committee discussed the possibility of purchasing a host site on the Asia-Pacific Computerized Information System (APCIS) and the development of an IOMOU web site.

(7) West and Central Africa region (Abuja MOU)

Established: 22 October 1999

Members: Angola, Benin, Cameroon, Cape Verde, Congo, Cote d'voire, Equatorial Guinea, Gabon, Gambia, Ghana, Guinea, Liberia, Mauritania, Namibia, Nigeria, Senegal, Sierra Leone, South Africa and Togo

(8) Black Sea region (Black Sea MOU)

Established: 7 April 2000

Members: Bulgaria, Georgia, Romania, the Russian Federation, Turkey and

Ukraine

(9) Arab States of the Gulf (Riyadh MOU)

Established: 30 June 2004

Members: The Kingdom of Bahrain, State of Kuwait, Sultanate of Oman, State of Qatar, Kingdom of Saudi Arabia, The United Arab Emirates

2.2.2 U.S.C.G.

In the 1970's, the U.S. Coast Guard increased its emphasis on the examination of foreign vessels. Although this emphasis was primarily driven by requirements to ensure compliance with the then new U.S. pollution prevention and navigation safety regulations, boarding officers also exercised Port State authority when instances of non-compliance with SOLAS and MARPOL were noted. In 1994, the U.S. introduced risk-management methodologies into the Port State Control program in order to allocate limited inspection resources to where they could do the most good, by identifying those ships, ship owners, classification societies and flag Administrations that were most often found lacking in meeting their international Convention responsibilities. In 2002, the classification society performance criteria were changed recognizing that the current method of comparing individual classification society performance against the average performance was becoming overly restrictive. Keeping in mind that demonstrated sustained high performance is the goal, the U.S.C.G. modified the methodology used to evaluate classification society performance by defining a fixed detention ratio level. Additionally, the U.S.C.G. has made significant strides towards the targeting of charterers. The Coast Guard is currently collecting information on charterers associated with detained vessels, and this information is being posted on the Port State Control Web site on the Internet. On 1 January 2001, the U.S.C.G. implemented an initiative to identify high-quality ships, so-called Qualship 21, quality shipping for the 21st century. The eligibility criteria for Qualship 21 were evaluated in these years. Recognizing that the current detention ratio that a flag state must meet is a moving target, as well as having reached a level commensurate with high quality performance, the detention ratio criterion was fixed in 2002. Beginning 1 August, 2002, U.S.C.G. enforced full compliance with the requirements of the STCW 95 Convention upon the ships entering U.S. ports. Although the U.S.A. is not a member of any MOU, the U.S.C.G. carry out effective PSC through co-operation with other MOUs.

2.2.3 Equasis

On 17 May 2000, the maritime administrations of France, Japan, United Kingdom, United States Coast Guard, Spain, Singapore and the European Commission signed the 'Memorandum of understanding on the establishment of the Equasis Information System' at the IMO Headquarters in London in conjunction with the 72nd Maritime Safety Committee meeting.

Equasis is a unique database collecting safety-related information on the world'merchant fleet from both public and private sources and making it easily accessible on the internet. (http://www.equasis.org/)It displays information from public authorities (Port State inspection and detention information from the three participating port State Control regions, i.e. Paris MOU, Tokyo MOU and the US Coast Guard) and industry players (such as information on class, insurance, participation in industry inspection schemes and quality organizations), all free of charge.

3

Targeting System & Boarding

3 - Targeting System & Boarding

In order to phase out substandard ships, Port State Control (PSC) has been intensified world-wide in the nineties to ensure compliance with applicable requirements for safety, pollution prevention and manning conditions on board. Substandard or non-compliant ships are detained and can not leave the port until reported serious/detainable deficiencies have been rectified. This of course implies possible delay and financial losses to ship operators.

PSC Inspectors are boarding a ship without announcement and primarily check the ship's documents for completeness and validity. If there are any grounds to believe that the ship is substantially not conforming with the international conventions, the inspector will carry out an expanded inspection of the ship's condition and the required equipment.

The Master will receive an official inspection report consisting of Form A and B. Form A lists the ship's details and the validity of the relevant certificates. Form B shows the list of deficiencies found (if any), with an action code which describes a timeframe for rectification for each deficiency. If clear grounds are established that the ship forms a hazard to safety and/or the environment, the PSCO has the right to detain the ship in port until the respective deficiencies have been rectified and resurveyed. The PSC authority will either resurvey by own inspectors or ask for a survey report from the Classification surveyor to verify the rectification. In case of a detention the PSC authority has the right to present a bill about their inspection activities. Any detention has to be reported as soon as possible by the authority to the flagstate, the classification society and IMO. The data about the inspection and the given timeframe for rectification are entered in a computer system used by all members of a regional PSC agreement. Most of the MOU accepted similar system with Paris MOU and USA Coast Guard targeting system. There for there is not necessity to explain all MOU's targeting system.

3.1 Paris MOU Targeting System

The port states authorities set overall percentage inspection rates to ensure that a minimum number of ships are inspected. They use ***targeting factors*** to focus inspection effort on those ships most likely to be substandard. Ships of a certain age and type are specifically selected for the purpose of conducting ***expanded inspections. Concentrated inspection campaigns*** are conducted to check on special matters or areas of concern.

Certain **selection criteria** such as the ship's flag, age and type, are believed to directly influence how well a ship is likely to be operated and in what condition a ship is likely to be found. To help PSCOs rank priority ships, the Paris MOU has developed a computerised targeting formula as part of its database system. This formula resulted in a **target factor (TF)** for each individual ship. By allocating points to each criteria a **scoring system** is employed and a ship is assigned a targeting factor. The target factor value of each ship is calculated in database (SIRENAC) on the basis of ship's profile and inspection history.

The targeting system consists of two modules:

a. the **generic factor** and
b. the **history factor.**

The generic factor for an individual ship is calculated by adding together several elements forming a generic profile of the ship. These include: flag states on **black list**, targeted ship type, non EU recognised classification society, age of the ship, above average class deficiency ratio and flag. The generic factor is updated when the particulars of the ship change or the status of its existing flag or class change. **The history factor** includes; ships which are entering the region for the first time in the last 12 months; or which have not been inspected in the last six months; ships with a pervious detention in the last 12 months, and those with a number of deficiencies during last 12 months. **The overall TF** is calculated by adding the generic and historic factor. The targeting factor is only a guideline for selecting ships. Each state or port may have its own priority list of ships to be inspected depending on the type of vessels visiting their ports.

Irrespective of the targeting factors and concentrated inspection campaigns there are a number of circumstances or **overriding factors** that would take a ship to the top of inspection list.

These include:
Ships that have been reported;
Ships reported as having outstanding deficiencies;
Where operational concerns about a ship exist;
Ships suspended from class.

If a ship has been inspected within the Paris MOU region during the previous six months and, on that occasion, was found to comply, the ship will in principle be exempted from further inspection, unless there are clear grounds to warrant further investigation. In selecting ships for inspection the Paris MOU Authorities will give priority to:

1. Ships visiting a port of a State, the Authority of which is a signatory to the Memorandum, for the first time or after an absence of 12 months or more. In the absence of appropriate data for this purpose, the Authorities will rely upon the available Sirenac data and inspect those ships which have not been registered in Sirenac following the entry into force of that database on 1 January 1993;
2. Ships not inspected by any Authority within the previous 6 months;

3. Ships whose statutory certificates on the ship' construction and equipment, issued in accordance with the Conventions, and the classification certificates, have been issued by an organization which is not recognized by the Authority;
4. Ships flying the flag of a State appearing in the black-list as published in the annual report of the MOU;
5. Ships which have been permitted by the Authority to leave a port of its State on certain conditions:
   a) deficiency to be rectified before departure
   b) deficiency to be rectified at the next port
   c) deficiencies to be rectified within 14 days
   d) deficiencies for which other conditions have been specified
   e) if ship related action has been taken and all deficiencies have been rectified;
6. Ships for which deficiencies have been recorded during a previous inspection, according to the number of deficiencies;
7. Ships which have been detained in a previous port;
8. Ships flying the flag of a non-Party to a relevant instrument;
9. Ships with class deficiency ratio above average;
10. Ships which are in a category for which expanded inspection has been decided;
    a) Oil Tankers, 5 years or less from the date of phasing out in accordance with Regulation 13G of Annex I to MARPOL 73/78
    b) Bulk Carriers, older than 12 years of age
    c) Passenger ships
    d) Gas and Chemical Tankers older than 10 years of age
11. Other ships above 13 years old.

3.1.1 Scoring System

Calculation of the Target Factor:

a. Generic Factor

The Generic Factor for an individual ship is calculated by adding together the applicable elements of its profile according to the elements below:

Targeted flag

A flag whose number of detentions in the last three years exceed its allowable limit based on a fixed yardstick (=7%). Graduated by increasing yardstick in steps of 3%. For example "medium to high risk" means detentions exceeded allowable limit using a yardstick of 10%.

(for detention % ref. Paris MOU Annual report)

Medium risk	(yardstick + 3 %)	TF + 4
Medium to High risk	(yardstick + 6 %)	TF + 8
High risk	(yardstick + 9 %)	TF + 14
Very High risk	(yardstick + 12 %)	TF + 20
Targeted ship type		TF + 5

i. Bulk carrier more than 12 years old.

ii. Gas Carrier more than 10 years old.

iii. Chemical Tanker more than 10 years old.

iv. Oil tanker GT〉3000 and more than 15 years old.

v. Passenger ship/ro-ro ferry more than 15 years old (other than ro-ro ferries and HS passenger craft operating in regular service under the provision of Council Dir. 1999/35/EC)

Non - EU recognised classification society TF + 3

A class society not appearing on the list of recognised societies published by EC Commission. If no class is recorded in the database (other than withdrawal or suspension of class for safety reasons) the ship will be assumed to be classed with an EU recognised class society.

Ships more than 12 years old

Graduated for non-targeted ship types (ref. above) and passenger ships

Age:

〉25 years TF + 3

21-24 TF + 2

13-20 TF + 1

Flag State has not ratified all conventions TF + 1

Flag states who have not ratified all main conventions.

(Ref. Relevant instruments in Paris Memorandum text, ratification information can be found on www.imo.org & www.ilo.org)

Targeted Class

Class with a 3-yr average record of detentions above the average class detention value using the excess of average rate as yardstick. A classification society whose number of detentions with class related deficiencies in the last three years exceeds the average class detention rate. Graduated by increasing the "excess of average"in steps of 2%. E.g. the overall class detention rate is 2.1% and the detention rate of a classification society is 4.1%, the "excess of average" value is 2%. (for detention % ref. Paris MOU Blue Book)

≤0% TF 0

0% - 2% TF + 1

〉2% - 4% TF + 2

〉4% TF + 3

The Generic Factor is updated when the particulars of the ship change or the status of its existing flag or class change.

b. History Factor

The History Factor is applied to the Generic Factor to reflect the actual condition of the ship found by inspection.

The History Factor is calculated by applying the elements below to each Paris MOU inspection of the ship carried out in the previous 12 months

Entering a region port for the first time in the last 12 months	TF + 20
No inspection recorded in the database in the last 12 months.	
Not inspected in last 6 months	TF + 10
No inspection recorded in the database in the last 6 months.	
Detained	TF + 15
Number of deficiencies:	
0	TF − 15
1 to 5	TF 0
6 to 10	TF + 5
11 to 20	TF + 10
21+	TF + 15

Outstanding deficiencies from last inspection

The value for the outstanding deficiencies is applied only in respect of the latest inspection.

- for each listed action taken "rectify deficiency at next port" or "Master instructed to rectify deficiency before departure" and for every two listed action taken "rectify deficiency within 14 days" and/or "other" TF + 1
- in case "all deficiencies rectified" is noted on the report TF − 2

The History Factor is updated at the end of each day.

Overall Target Factor is calculated by adding the Generic and History Factor but cannot be less than the Generic Factor. The overall Target Factors are re-calculated at the end of each day.

The TF is in use within the Paris MOU on PSC as a tool for selecting ships eligible for an inspection only. The TF is not an indication of the quality of the ship.

Example

Date	: 20-07-2002		
Vessel	: Paris MOU		
Flag	: Honduras		
Type	: Passenger ship		
Class	: Honduras International Naval Survey&Inspection Bureau		
Year build	: 1958		
Honduras detention %	: 27.71%		
Average Class detention %	: 2.1%		
HINSIB detention %	: 13.04%		
Inspection history	:		
Ghent Belgium		10-07-2001	9 Deficiencies
Rotterdam The Netherlands		11-09-2001	5 Deficiencies
Hamburg Germany		15-01-2002	16 Deficiencies
Genoa Italy		04-07-2002	20 Deficiencies
		−8 Rectify before departure	
		−10 Rectify within 14 days	
		−2 Other	

Generic Factor

Targeted flag	Very High Risk	+ 20
Targeted ship type	Passenger ship	+ 5
Non-EU recognised classif. society	HINSIB	+ 3
Ships more than 12 years old		+ 3

FS has not ratified all conventions	LL Pro.88, SOLAS Pro.88, ILO147	+ 1
Targeted Class	Excess of average 10.94%	+ 3
		35

History Factor

Entering a region port for the first time in the last 12 months		0
Not inspected in last 6 months		0
Detained		0
Number of deficiencies	Rotterdam	0
	Hamburg	+ 10
	Genova	+ 10
Outstanding deficiencies from last inspection		
"Master instructed to rectify deficiency before departure"		+ 8
"rectify deficiency within 14 days"		+ 5
"other (specify in clear text)"		+ 1
		34

Overall Target Factor

Generic Factor	: 35
History Factor	: 34
Overall Target Factor	: 69

3.2 USCG Targeting System (Boarding Priority Matrix)

United States does not take part in any of the regional agreements on PSC. It undertakes control measures on a unilateral basis. On May 1, 1994 the US Coast Guard introduced its revised PSC initiative. The primary objective of this program is to identify high risk foreign merchant ships based on the performance records of their owners, operators, classification societies and flag states; and to systematically target ships for boarding.

In the USA, there is no agreement or MOU which is specifically dedicated to

PSC. Therefore it is not possible to have a conclusive list of conventions enforced by the US Coast Guard under the PSC program. The US exercises its PSC authority through the US Coast Guard's long standing foreign vessel boarding program, nor referred to as PSC Program.

In US PSC system the following have been considered as the primary factors in making the decision to conduct an onboard inspection:

- Owners and operators' list;
- Classification societies;
- Flag states;
- The boarding history of the vessel;
- Vessel type;

According to US Coast Guard the first three entities directly influence a vessel's operational condition and compliance with international safety and environmental protection standards. Therefore, if any of these entities fails to undertake its responsibilities fully for a ship's safe operation, then the ship is likely to be considered a substandard vessel by the US Coast Guard.

Boarding Priority Matrix

The possibility of a foreign vessel being boarded is based on the number of points the vessel acquires under the boarding priority matrix. However, the points assigned to a vessel under this targeting regime do not classify the vessel as substandard; only a boarding and examination can reveal such conditions.

The boarding matrix is used to decide which ships PSC officers should board on any given day, in any given port.

Owner	Flag	Class	History	Ship Type
5 points Listed owner or operator	**7 points** Listed Flag State	**Priority I** Ten arrivals with detention ratio more than 4 times the average OR Ten arrivals and involved with at least one detention in the previous 3 years.	**5 points each** Detention within the previous 12 months.	**1 point** Oil or chemical tanker.
		5 points Ten arrivals with a detention ratio between 3&4 times the average.	**1 point each** Other operational control within the previous 12 months.	**1 point** Gas Carrier
		3 points Ten arrivals with a detention ratio between 2&3 times the average.	**1 point each** Casualty within the previous 12 months.	**2 points** Bulk Freighter over 10 years old.
		1 point Ten arrivals with a detention ratio between the average and twice the average.	**1 point each** Violation within the previous 12 months.	**1 point** Passenger Ship
		0 points Ten arrivals with a detention ratio below the average OR Ten arrivals with no detentions in the previous 3 years.	**1 point each** Not boarded within the previous 6 months.	**2 points** Carrying low value commodities in bulk.

Boarding Priorities

a. Priority I Vessels:

Priority I vessels include:

- 17 or more points on the Matrix, or

- Ships involved in a marine casualty that may have affected seaworthiness, or
- USCG Captain of the Port determines a vessel to be a potential hazard to the port or the environment, or
- Ships whose classification society has ten or more arrivals the previous year and a detention ratio more than four times the average, or
- Ships whose classification society has less than ten arrivals the pervious year and have been associated with at least one detention.

Priority I vessels are targeted for examination prior to entry into US ports. Port entry of these vessels may be restricted until they are examined by the Coast Guard. Where feasible these vessels are boarded prior to port entry to ensure deficiencies are corrected. Otherwise, they are boarded upon entry and prior to commencement of cargo transfer operations or passenger embarkation.

b. Priority II Vessels

Priority II vessels include:

- 7 to 16 points on the Matrix, or
- Outstanding requirements from a previous boarding in this or another US port, or the vessel is overdue for an annual tank or passenger exam.

Priority II vessels are targeted for boarding prior to commencement of cargo transfer operations or passenger embarkation. Cargo operations of these vessels may be restricted until the vessels are examined by the Coast Guard. An exemption to the requirement for boarding prior to commencement of cargo transfer operations or passenger embarkation may be granted if there are clear indications that the ship is in substantial compliance with applicable standards.

c. Priority III Vessels

Priority III vessels include:

- 4 to 6 points on the Matrix, or

- Alleged deficiencies reported, or
- The vessel is overdue for an annual examination, or quarterly passenger re-exam.

Priority III vessels may be targeted for boarding after entry into port, but there are no restrictions on commencement of cargo transfer operations or passenger embarkation.

d. Priority IV Vessels

Priority IV vessels include:

- 3 or fewer points on the Matrix.

Priority IV vessels are not targeted for boarding, but may be boarded and examined by the US Coast Guard at the discretion of the local Captain of the Port or Officer in Charge, Marine Inspection.

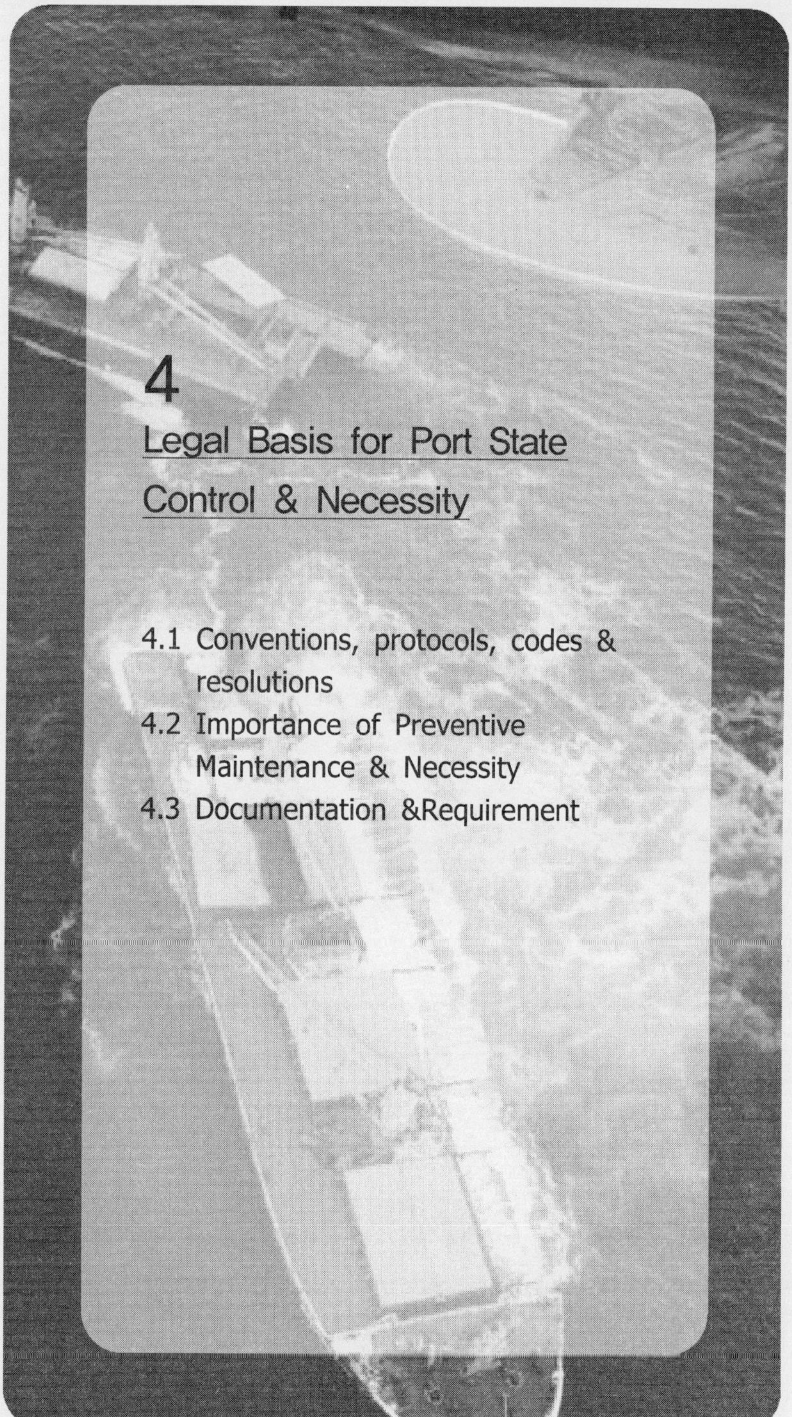

4

Legal Basis for Port State Control & Necessity

4 - Legal Basis for Port State Control & Necessity

4.1 Conventions, protocols, codes & resolutions

Port State Control inspections (PSC) is exercised on the legal basis specified in applicable regulations specified in the following International instruments:

- SOLAS
- LOAD LINE
- MARPOL
- STCW
- COLLREG
- TONNAGE
- ISM
- ILO 147
- BCH CODE
- IGC CODE
- IBC CODE
- IMO Res. A.787(19)

Regulation 19 specify the control procedure in SOLAS
Article 21 " " " " " Load Lines 66 Convention.
" 5 " inspection procedures of ships
" X and Regulation I/4 specify the control procedures of STCW
Furthermore, IMO resolution A.787(19), adopted on 23 November 1995, describes in more details "Procedures for Port State Control".
To exercise control of the provisions of the Codes is covered by SOLAS.
An overview of mandatory certificates and mandatory documentation are listed in chapter 4.3.

4.2 Importance of Preventive Maintenance & Necessity

Due to the operation of the ship and exposure to environmental conditions, such as air, humidity, heat and seawater, the ship and its equipment deteriorate. Further, as the time passes by, certain documentation, services and equipment become invalid, out-dated or non-compliant.

This makes maintenance necessary in order to maintain compliance with the applicable requirements on safety and pollution prevention. Operators carry out maintenance either as corrective maintenance or preventive maintenance.

Corrective maintenance means taking corrective actions **after** deficiencies have occurred. Corrective maintenance is an **unacceptabl**e approach as it implies that at times deficiencies can be found on board compromising the safety of the ship, its personnel and the environment. When serious deficiencies are found during class surveys, Flag State inspections or Port State inspection, the ship may be detained until rectification.

Preventive maintenance means taking preventive actions **before** deficiencies occur. This means continuous compliance, no deficiencies on board and therefore no ground for detention.

Every Company, master and/or responsible officers should remind themselves of the provisions of SOLAS Regulation I/11 "Maintenance of condition after survey", which states that;

The condition of the ship and its equipment shall be maintained to confirm with the provisions of the present regulations to ensure that the ship in all respects will remain fit to proceed to sea without danger to the ship or persons on board.

This means that the ship shall in all respect be kept to the same standard as it was when inspected and the applicable certificate(s) issued. Necessary maintenance of the ship to keep its standards are necessary, otherwise the

requirements of SOLAS Reg. I/11 is not complied with. Lack of maintenance may result in deterioration of the ship's standards to a degree that it is considered by the PSCO "not in all respect fit to proceed to sea". This may be considered clear grounds for **the PSCO on the basis of the provisions of SOLAS Regulation I/11, for more detailed inspections or detention.** Otherwise clear grounds means also that the PSCO has found that the ship, its equipment, or its crew does not corresponds substantially with the provisions of the relevant conventions. Also when the master or crew members are not familiar with essential shipboard procedures relating to the safety of the ship or prevention of pollution, this may be considered as clear grounds. Even though the condition is that the ship's standards shall correspond to the applicable provisions of Conventions after survey, normal wear and tear resulting in deterioration of the overall standards pending on ships age, should be kept in mind when inspections are carried out. However, the master and/or responsible officer(s) should always maintain the ship's condition to a highest possible maintenance level to be prepared for possible inspections also by PSCO. They should consider whether any inspection items could be a problem for the ship concerned, and if that is the case take necessary steps to improve the condition so that the standards correspond to the particular provisions of applicable Conventions.

4.3 Documentation &Requirement

1 International Tonnage Certificate (1969) (Int. Tonnage Conv. 1969)
2 Passengers Ship Safety Certificate (SOLAS 74/Ch. I/7)
3 Cargo Ship Safety Construction Certificate (SOLAS 74/CH. I/10)
4 Cargo Ship Safety Equipment Certificate (SOLAS 74/Ch. I/8)
5 Cargo Ship Safety Radio Radiotelegraphy Certificate (SOLAS 74/CH. IV)
6 Cargo Ship Safety Radio telephony Certificate "
7 Cargo Ship Safety Certificate (Comb. Cert. SOLAS Ch. I/8-9-10)
8 Exemption Certificate)(SOLAS 74/Ch.I/4)
9 Document of Compliance)(SOLAS 74 R II-2/54)

10 Trading Certificate or Trading Permit (National Req. Pending Flag)

11 International Certificate of Fitness for the Carriage of Liquefied Gases in Bulk; Certificate of Fitness for the Carriage of Liquefied Gases in Bulk (SOLAS Ch. VII, IGC-Code)

12 International Certificate of Fitness for the Carriage of Dangerous Chemical in bulk; Certificate of Fitness for the Carriage of Dangerous Chemicals in Bulk or the ships P & A Manual(SOLAS 7/ MARPOL 73/78 – Annex II/IBC-Code)

13 International Oil Prevention Certificate(MARPOL 73/78 Annex II/5)

14 International Pollution Prevention Certificate for the Carriage of Noxious Liquid Substances in bulk(MARPOL 73/78 Annex II/11)

15 International Load Line Certificate (1966); International Load Line Exemption Certificate(Int. LL Conv. Article 16)

16 Oil Record Book part I(MARPOL Annex I/20)

17 Oil Record Book part II (MARPOL 73/78 Annex I/20

18 Oil Record Book Chemicals (MARPOL 73/78 Annex II/9)

19 Cargo Record Book(MARPOL Annex II/9)

20 Cargo Gear Record Book(ILO Conv. 134 Article 4.3e)

21 Minimum Safe Manning Document; **Certificates of Competency**(SOLAS74 Ch.V/13)

22 Medical certificates, Cf. ILO Conv. No 73 concerning Medical examination of seafarers

23 Approved Stability information (SOLAS 74 Ch. II-1/22)

24 Approved Shipboard Oil Pollution Emergency Plan(MARPOL Annex I/26)

25 Safety Management Certificate (1 July 1998) pending ship type(SOLAS Ch. IX/4 – ISM – Code)

26 Classification Certificate of the hull, electrical and machinery installations issued by the classification society if the ship is classed (SOLAS 74 Ch. I/7 and I/10)

27 Survey Report Filed(in case of bulk carriers or oil tankers)

28 Report of previous Port State Control inspections

29 For ro-ro passenger ships, information on A/Amax ratio Ro/Ro Max-Rat

30 Valid Certificate for inflatable life-rafts and Free-Float release equipment (SOLAS Ch.III/19.8 and III/19.9)

31 Valid Certificate for fixed fire-fighting equipment. (SOLAS Ch. I/7& I/10)

32 Garbage Record Book(MARPOL Annex V)

33 Dangerous goods manifest or stowage plan (SOLAS VII/5(5) MARPOL Annex III/4)

34 Document of authorizations for the carriage of grain (SOLAS VI/9)

35 Certificate of insurance or other financial security in respect of civil liability For pollution damage(CLC 69, art VII)

36 Approved cargo securing manual(SOLAS VI/5.6 and VII/5.6)

If the ship does not have the relevant valid documentation's (certificates), it would be a criteria for a detainable deficiency or clear grounds for more detailed inspection. In addition to the general control of above listed certificate and documents, examinations/inspections of the following will normally be given priority by PSCO:

Nautical publication (SOLAS 74 R V/20)

Navigational equipment (SOLAS 74 R V/12 and 19)

Emergency starting and running tests (SOLAS 74 R II-2 - 4.3)

Lifesaving equipment. Rafts FF (SOLAS 74 R III/20, 23, 26 and 29)

Emergency Generator (start/stop only) (SOLAS 74 R II-1/42&43)

Hull corrosion and damages (Load Lines) (SOLAS 74 R I/11)

Main engine & aux. engines (SOLAS 74 R II/26, 27 &28)

Oily water separator 15 ppm alarm (MARPOL Annex I/16(1))

Oil discharge monitor (ODM) (MARPOL Annex I/16)

Charts corrected and proper scale (SOLAS 74 R V/20)

Fire safety Control plan (SOLAS 74 R II-2/20)

Ventilation inlets/outlets (SOLAS 74 R II-2/16.9 & 48)

Emergency training and drills (Log book rec. SOLAS 74 R III/18)

Emergency lighting/batteries (SOLAS 74 R II/42 &43)

Deck- and hatches corrosion and damages (LL 1966)

Steering gear – incl. auxiliary & emergency (Bridge inspection only – SOLAS 74 R V/19)

Cleanliness in engine room (SOLAS 74 R II-1/26 and ILO 134)

Cleanliness in accommodation (ILO 92 & 133)

5
Port State Control Inspection

5 - Port State Control Inspection

Every ship should be prepared for survey and/or inspection by a Port State Control Officer (PSCO). The PSCO shall prior to any Port State Control inspection follow applicable procedure by introducing himself to the master and/or the responsible officers on board prior to the control. Every PSCO shall be duly qualified to conduct Port State Control inspections.

Port State control is carried out by a Port State Control Officer (PSCO). The PSCO is a properly qualified person, authorized to carry out port State control inspections in accordance with the IMO conventions, by the maritime authority of the port State and acts under its responsibility. All PSCOs carry an identity card, issued by their maritime authorities. The PSCO should be an experienced person qualified as a flag State surveyor and able to communicate with the master and key crew members in English.

He should have no commercial interest in the port, the ship or be employed by or on behalf of a classification society.

The PSCO is issued with an identity card as evidence of his authority to carry out inspections. All PSCOs should also carry a copy of the General Procedural Guidelines for PSCOs from IMO resolution A.787(19) for ready reference when carrying out inspections.

Port State Control Inspections may be conducted on the following basis:

1 initiative of the Port State Administration;

2 the request of, or on the basis of, information regarding a ship provided by another Administration

3 information regarding a ship provided by a member of the crew, a professional body, an association, a trade union or any other individual with

an interest in the safety of the ship, its crew and passengers, or the protection of the marine environment.

PSC inspections may be on random, targeted or periodical basis. The following types of PSC inspections are used in PSC:

1. **Initial Inspection (random)**
2. **More detailed inspection (escalated)**
3. **Expanded inspection (targeted/periodical)**

5.1 Initial Inspection

The PSCO will normally examine the vessels relevant certificates and documents etc. and the overall condition of the ship.

The certificates and documents listed previous chapter should therefore be readily available and presented to the PSCO at his request.

5.2 More detailed inspection

An inspection conducted when there are **clear grounds** for believing that the condition of the ship, its equipment, or its crew does not comply with the requirements of the relevant conventions. The inspection may focus on one area or be across various areas. The following may be considered to be clear grounds for more detailed inspections by a PSCO:

1. the absence of principle equipment or arrangements required by the conventions;
2. evidence from the review of the ship's certificates that a certificate or certificates are clearly invalid;
3. evidence that the ship's logs, manuals or other documentation are not on board, are not maintained, or are falsely maintained;
4. in the opinion of the PSCO's general impression and observation serious hull

or structural deterioration or deficiencies exist that may place at risk the structural, watertight or weathertight integrity of the ship;

5. in the opinion of the PSCO's general impression or observation serious deficiencies exist in the safety, pollution prevention, or navigational equipment;
6. information or evidence that the master or crew is not familiar with essential shipboard operations relating to safety of ships or the prevention of pollution, or that such operations have not been carried out
7. indication that key crew members may not be able to communicate with each other or with other persons on board;
8. absence of an up-to-date muster list, fire control plan, and for passenger ships, a damage control plan;
9. the emission of false distress alerts not followed by proper cancellation procedures;
10. receipt of a report or complaint containing information that a ship appears to be substandard;
11. the ship has been reported by pilots or port authorities or others as having deficiencies which may prejudice safe navigation;
12. ships whose statutory certificates have been issued by an organizations which is not recognised under the term of Council Directive 94/57/EC of 22 November 1994 on common rules and standards for ship inspection and survey organizations and for the relevant activities of maritime Administration;
13. the ship has been involved in a collision on its way to the port;
14. the ship is in a category for which expanded inspection has been decided;
15. the ship has been suspended from their class for safety reasons in the preceding six months;
16. the ship has been accused of an alleged violation of the provisions on discharge of harmful substances or effluents

17. the ship's statutory certificates on the ship's construction and equipment, have been issued by an organizations which is not recognised by the Authority
18. the ship flying the flag of a State appearing in the three-year rolling average table of above average detentions in the annual report of MOU.

5.3 Expanded inspection

An inspection conducted according to non-mandatory guidelines only once during 12 months period for certain types of ships (tankers, bulkers and passenger ships) and certain categories of age and size.

Oil tankers, bulk carriers, gas and chemical carriers and passenger ships are subject to expanded inspections once during a period of 12 months. These inspections could be carried out in accordance with provisions stated below:

- Oil tankers, five years or less from the date of phasing out in accordance with MARPOL 73/78, Annex I Regulation 13 G, i. e.
- a crude oil tanker of 20.000 DWT and above or a product carrier of 30.000 DWT and above, not meeting the requirements of a new oil tanker as defined in Regulation 1 (26) of Annex I of MARPOL 73/78, will be subject to expanded inspection:
- 20 years after its date of delivery as indicated on the Supplement, Form B, to the IOPP Certificate, or
- 25 years after that date, if the ship's wing tanks or double bottom space not used for carriage of oil meet the requirements of Regulation 13 G (4) of the Annex, unless it has been reconstructed to comply with Regulation 13 F of the same Annex.
- an oil tanker as mentioned above meeting the requirements of a new oil tanker as defined in Regulation 1 (26) of Annex I to MARPOL 73/78 will be subject to expanded inspection:

- 25 years after its date of delivery as indicated on the Supplement, Form B, to the IOPP Certificate, unless it complies with or has been reconstructed to comply with Regulation 13 F of the Annex.
- Bulk carrier, older than 12 years of age, as determined on the basis of the date of construction indicated in the ship's safety certificate. **Such expanded inspection will be conducted only ones during a period of 12 months by any of the competent authorities of the MOU.**
- Gas and chemical tankers older than 10 years of age, as determined on the basis of construction indicated in the ship's safety certificate
- In case of passenger ship operating on a regular schedule in or out of a port in an EU member state, the competent authority of the Member State shall carry out an expanded inspection of each ship. When a passenger ship operates such a schedule between ports in Member States, one of the States between which the ship is operating shall undertake the expanded inspection. To the extent it is applicable, the following examinations may be considered as part of an expanded inspection. However, when the examinations are carried out, the master and/or responsible officers should remind the PSCO that it may jeopardize the safe execution of certain on-board operations, e. g. cargo operation, if the tests having a direct effect thereon, and are required to be executed during such operations:
  a. execution of black-out and start of emergency generator;
  b. inspection of emergency lighting and back up sources including batteries;
  c. operation of emergency fire-pump with two firehouses connected to the main fire line;
  d. operation of bilge pumps;
  e. closing of watertight doors;
  f. lowering of a seaside lifeboat to the water level and test the release mechanism;
  g. inspection of fire dampers to engine room, cargo holds and accommodation;

h. test of remote emergency stop e. g. boiler, ventilation and fuel pumps;

i. testing of steering gear including auxiliary steering gear;

j. inspection and testing of emergency source of power to radio installations;

k. inspection and, to the extent possible, test of engine-room separators;

Additional expanded inspections, which might be carried out for oil tankers:

- fixed-deck foam system;
- fire-fighting equipment in general;
- inspections of fire dampers to pump room;
- Control of pressure of inert gas and oxygen content thereof; check of survey report file to identify possible suspect areas requiring inspections.

Additional expanded inspections, which might be carried out for bulk carriers:

- corrosion of deck machinery foundations
- deformation and/or corrosion of hatch covers
- cracks and/or local corrosion in transverse bulkheads
- access to cargo holds
- check of Survey Report File to identify possible suspect areas requiring further inspections

Additional expanded inspections, which might be carried out on gas and chemical carriers:

- cargo tank monitoring and safety devices relating to temperature, pressure, gas detection, and ullage
- oxygen analysing and explosimeter devices, inc. their calibration
- availability of chemical detection equipment (bellows) with an appropriate number of suitable gas detection tubes for the specific cargoes being carried
- cabin escape sets giving suitable respiratory and eye protection, for every person

- onboard (if required by products listed on International Certificate of Fitness or Certificate of Fitness for the Carriage of Dangerous Chemicals in Bulk or Liquefied Gases in Bulk as Applicable)
- check that the product(s) being carried is listed in the International Certificate of Fitness or Certificate of Fitness for the Carriage of Dangerous Chemicals in Bulk or Liquefied Gases in Bulk as applicable
- the fixed fire fighting installations on deck whether they be foam or dry chemical or other as required by the product(s) carried

Additional expanded inspections, which might be carried out on passenger ships;

- testing of fire detection and alarm system
- testing of proper closing of fire doors
- testing of public address system
- fire drill where, as a minimum, all sets of fireman's outfits must be demonstrated and part of the catering staff shall take part
- demonstration that key crew members are acquainted with the damage control plan

As mentioned above a PSCO might always find "clear grounds" for carrying out more detailed inspections. "**Clear grounds**" exist when the inspector finds evidence which in his professional judgement warrant a more detailed inspection of the ship, its equipment or its crew. During an expanded inspection the PSCO might find evidence which in his judgement warrants a more detailed inspection - "**Clear grounds**". Whenever there are clear grounds for believing that the condition of a ship or its equipment or crew does not substantially meet the relevant requirements of a Convention, a more detailed inspection shall be carried out, including further examination of compliance with on-board operational requirements. Therefore the master and/or responsible officers should bear in mind that en expanded inspection could be continued by a more detailed inspection, concentrating on an area which in the opinion of the PSCO need further examination, and might comprise of the following priorities;

- navigational safety
- communication
- fire safety
- life-saving appliances etc

5.4 Port State Control & Cargo Operations

There are areas, where deficiencies do not warrant a detention of a ship, but might lead to suspension of cargo operation. Deficiencies such as failure of the proper operation of inert gas system, improper operation of cargo related gear or machinery, crude oil washing procedures might be considered as grounds for such suspensions.

Inert gas system

If an examination/inspection of the inert gas system is carried out, it may include examination of instrumentation fitted for continuous monitoring. Such monitoring includes indicating and permanently recording at all times when inert gas is being supplied, the pressure and the oxygen content of the gas in the inert gas supply main. Reference to the permanent recorder must indicate if the system had been operating before and during the cargo discharge in a satisfactory manner. If conditions specified in the COW Operation and Equipment Manual are not being met, then the washing will be stopped until satisfactory conditions are restored. A further precautionary measure is that the oxygen level in each tank to be washed, is to be determined at the tank. The explosivemeters must have been calibrated and will be inspected to ensure that they are in good working order. Readings from tanks already washed in port prior to inspection should be available for examination, and spot checks on readings might be instituted

Crude oil washing.

Normally the PSCO will ensure that all crude carriers either required to have crude oil washing system or where the owner or operator chooses to install a crude oil washing system in order to comply with Regulation 13 of Annex I of MARPOL 73/78. In addition, compliance will be ensured with the operational requirements set out in the revised Specifications of the Design Operation and Control of Crude Oil Washing Systems (IMO Res. A.446(XI), as amended by IMO Res. A.497(XII)) This is done in the ports where the cargo is unloaded. Further, the inspection may cover the entire operation of crude oil washing or only certain aspects of it. It is; however, in the interest of all concerned that the ship's records with regard to the crude oil washing operations are maintained at all times so that a PSCO may verify those operations undertaken prior to the applicable inspection. It will be determined from the ship's records that the pre-crude oil wash operation has been carried out and that all instruments functioned correctly. If a tanker is engaged in multiple port discharge, the Oil Record Book (ORB) should indicate if tanks have been crude oil washed at previous discharge ports or at sea. It will be determined that all tanks which will or may be used to contain ballast on the forthcoming voyage will be crude oil washed before the ship departs from the port. There is no obligation to wash any tank other than ballast tanks at a discharge port except that each of these other tanks must be washed at least in accordance with paragraph 6.1 of the revised Specifications (IMORes. A.446(XI) as amended by Res.A.497(XII). The Oil Record Book (ORB) will be examined to verify this has been complied with. All crude oil washing must be completed before a ship leaves its final port of discharge. If tanks are not being washed in one of the preferred orders given in the COW Operations and Equipment Manual, the PSCO will determine the reasons for this, and decide whether the order of tank washing are acceptable. For each tank being washed it will be ensured that the operation is in accordance with the COW Operations and Equipment Manual and that;

- the deck mounted machines and the submerged machines are operating either by reference to indicators, the sound patterns or other approved methods

- the deck mounted machines, where applicable, are programmed as stated
- the duration of wash is as required, and
- the number of tank washing machines being used simultaneously does not exceed that specified.

All tanks that have been crude oil washed are to be stripped. It will be ascertained that the adequacy of stripping has been checked or will be checked before the ship leaves its final port of discharge.

Tanks that have been crude oil washed at sea should be recorded in the ORB. These tanks must be left empty between discharge ports for inspections at the next discharge port. Tanks that are designated ballast tanks should be listed in the COW Operations and Equipment Manual. It is, however, left to the discretion of the master and/or responsible officer to decide which tanks may be used for ballast on the forthcoming voyage. It should be determined from the ORB that additional ballast water has not been put into tanks, which have not been crude oil washed during previous voyages. It will be verified that the departure ballast tanks are stripped as complete as possible. The methods to avoid vapour emission where locally required should be provided in the COW Operations and Equipment Manual and they must be adhered to. The PSCO will ensure that this has been complied with When departure ballast has to be shifted, the discharge into the sea must be in compliance with Regulation 9 of Annex I to MARPOL 73/78. The ORB will be examined to ensure compliance with this.

5.5 Detention reasons

A PSCO may impose the following courses of action on a ship:

a. Rectification of deficiencies prior to departure;

b. Rectification of deficiencies in the next port, under specific conditions;

c. Rectification of (minor) deficiencies (only) within 14 days;

d. Detention of the ship.

Following an inspection the PSCO has to decide which action has to be taken to correct deficiencies found and the time within which the corrections are to be made. If the deficiencies found are serious the PSCO has to decide whether he should prevent the ship from sailing until they are rectified.

In principle, all deficiencies must be rectified before departure of the ship. It is up to the professional judgement of the PSCO to decide that he has to board the ship on a second occasion to check personally if all deficiencies have indeed been rectified.

The following are the main criteria for the detention of a ship:

a ship which is unsafe to proceed to sea will be detained upon the first inspection, irrespective of the time the ship is scheduled to stay in port;

the deficiencies on a ship are so serious that they will have to be rectified before the ship sails.

In case deficiencies are clearly hazardous to safety, health or the environment, the maritime authorities will ensure that the hazard is rectified before the ship is allowed to proceed to sea and for this purpose they will either detain the vessel or issue a formal prohibition of a ship to continue an operation. The flag State will be notified as soon as possible.

If deficiencies cannot be remedied in the port of inspection, the maritime authority may allow the ship to proceed to another port, subject to any appropriate conditions determined by the maritime authority of the port of departure, with a view to ensuring that the ship can so proceed without unreasonable danger to safety, health or the environment. In this case a follow-up inspection will normally be carried out in this respective port.

In the event of a detention of a ship, the PSCO will note information on the owner or operator of the vessel at the time of the detention. The master will be asked to sign to confirm this information.

The port state should not charge the ship for any general inspection. However, inspection charges can be expected if the ship invites a port state to undertake inspection or if the ship is detained and the PSCO has to return to the vessel for a re-inspection. That's why it is important that the ship should ensure that all deficiencies are properly rectified before asking for a re-inspection. It is also possible that there may be charges if there are overriding factors.

In case of a ship arrest, the claimant needs to satisfy various legal tests before a hearing in the local court. In case of an appeal the courts usually hear any appeal made by a shipowner who is suffering substantial losses from the arrest. Unlike the arrest of a ship, a PSC detention order can be issued at the sole discretion of a PSCO, without prior consideration of the merits by a judge. An unduly detained ship will be entitled to compensation and can appeal against the detention if the detained ship can prove the wrongful detention.

Under different jurisdictions, there are limited rights of appeal against a port state detention order. However, appeal against a detention order takes quite a long time and does not stop the process of detention.

There follows an exhaustive list of deficiencies, grouped under relevant instruments which are considered serious enough to warrant the detention of the ship.

Detention items for the ships:

a. General

Lack of valid certificates and documents as required by the relevant instruments: However, ships flying the flag of States not a party to a Convention (relevant instrument) or not having implemented another relevant instrument are not entitled to carry the certificates provided for by the Convention or other relevant instrument. Therefore, absence of the required certificates will not by itself constitute reason to detain these ships, however, in applying the 'no more favourable treatment' clause, substantial compliance with the provisions must be required before the ship sails.

b. Areas under SOLAS

(1) failure of proper operation of propulsion and other essential machinery, as well as electrical installations;

(2) insufficient cleanliness of engine room, excess amount of oily-water mixtures in bilges, insulation of piping including exhaust pipes in engine room contaminated by oil, improper operation of bilge pumping arrangements;

(3) failure of the proper operation of emergency generator, lighting, batteries and switches;

(4) failure of the proper operation of the main and auxiliary steering gear;

(5) absence, insufficient capacity or serious deterioration of personal lifesaving appliances, survival craft and launching arrangements;

(6) absence, non-compliance or substantial deterioration to the extent that it can not comply with its intended use of fire detection system, fire alarms, fire fighting equipment, fixed fire extinguishing installation, ventilation valves, fire dampers, quick closing devices;

(7) absence, substantial deterioration or failure of proper operation of the cargo deck area fire protection on tankers;

(8) absence, non-compliance or serious deterioration of lights, shapes or sound signals;

(9) absence or failure of the proper operation of the radio equipment for distress and safety communication;

(10) absence or failure of the proper operation of navigation equipment, taking the provisions of Regulation V/12(o) of SOLAS 74 into account;

(11) absence of corrected navigational charts, and/or all other relevant nautical publications necessary for the intended voyage, taking into

account that type approved electronic chart display and information system (ECDIS) operating on official data may be used as a substitute for the charts;

(12) absence of non-sparking exhaust ventilation for cargo pump rooms (Regulation II-2/59.3.1 of SOLAS 74);

(13) Serious deficiency in the operational requirements listed in 5.5 of this Annex;

(14) Number, composition or certification of crew not corresponding with safe manning document;

(15) Failure to carry out the enhanced survey programme SOLAS 74, Chapter XI, Regulation 2;

(16) Absence or failure of a VDR, when its use is compulsory.

c. Areas under the IBC Code

(1) transport of a substance not mentioned in the Certificate of Fitness or missing cargo information (16.2);

(2) missing or damaged high pressure safety devices (8.2.3);

(3) electrical installations not intrinsically safe or corresponding to code requirements (10.2.3);

(4) sources of ignition in hazardous locations referred to in 10.2 (11.3.15);

(5) contraventions of special requirements (15);

(6) exceeding of maximum allowable cargo quantity per tank (16.1);

(7) insufficient heat protection for sensitive products (16.6).

d. Areas under the IGC Code

(1) transport of a substance not mentioned in the Certificate of Fitness or missing cargo information (18.1);

(2) missing closing devices for accommodations or service spaces (3.2.6);

(3) bulkhead not gastight (3.3.2);

(4) defective air locks (3.6);

(5) missing or defective quick closing valves (5.6);

(6) missing or defective safety valves (8.2);

(7) electrical installations not intrinsically safe or not corresponding to code requirements (10.2.4);

(8) ventilators in cargo area not operable (12.1);

(9) pressure alarms for cargo tanks not operable (13.4.1);

(10) gas detection plant and/or toxic gas detection plant defective (13.6);

(11) transport of substances to be inhibited without valid inhibitor certificate (17/19).

e. Areas under LOADLINES

(1) significant areas of damage or corrosion, or pitting of plating and associated stiffening in decks and hull effecting seaworthiness or strength to take local loads, unless proper temporary repairs for a voyage to a port for permanent repairs have been carried out;

(2) a recognized case of insufficient stability;

(3) absence of sufficient and reliable information, in an approved form, which by rapid and simple means, enables the master to arrange for the loading and ballasting of his ship in such a way that a safe margin of stability is maintained at all stages and at varying conditions of the voyage, and that the creation of any unacceptable stresses in the ship's structure are avoided;

(4) absence, substantial deterioration or defective closing devices, hatch closing arrangements and water tight doors;

(5) overloading;

(6) absence of or impossibility to read draught mark.

f. Areas under Annex I to MARPOL

(1) absence, serious deterioration or failure of proper operation of the oily-water filtering equipment, the oil discharge monitoring and control system or the 15 ppm alarm arrangements;

(2) remaining capacity of slop and/or sludge tank insufficient for the intended voyage;

(3) oil record book not available (20(5));

(4) unauthorized discharge bypass fitted;

(5) survey report file missing or not in conformity with Regulation 13G(3)(b) of the Convention.

g. Areas under Annex II to MARPOL

(1) absence of the P&A Manual;

(2) cargo is not categorized (3(4));

(3) no cargo record book available (9(6));

(4) transport of oil-like substances without satisfying the requirements (14);

(5) unauthorized discharge by-pass fitted.

h. Areas under Annex V to MARPOL

(1) absence of the garbage management plan;

(2) no garbage record book available;

(3) ship's personnel not familiar with disposal/discharge requirements of garbage management plan.

i. Areas under STCW

(1) Failure of seafarers to hold a certificate, to have an appropriate certificate, to have a valid dispensation or to provide documentary proof that an application for an endorsement has been submitted to the flag State Administration;

(2) Failure to comply with the applicable safe manning requirements of the flag State Administration;

(3) Failure of navigational or engineering watch arrangements to conform to the requirements specified for the ship by the flag State Administration;

(4) Absence in a watch of a person qualified to operate equipment essential to safe navigation, safety radio communications or the prevention of marine pollution;

(5) Failure to provide proof of professional proficiency for the duties assigned to seafarers for the safety of the ship and the prevention of pollution;

(6) Inability to provide for the first watch at the commencement of a voyage and for subsequent relieving watches persons who are sufficiently rested and otherwise fit for duty.

j. Areas under ILO Conventions

(1) insufficient food for voyage to next port;

(2) insufficient potable water for voyage to next port;

(3) excessively unsanitary conditions on board;

(4) no heating in accommodation of a ship operating in areas where temperatures maybe excessively low;

(5) excessive garbage, blockage by equipment or cargo or otherwise unsafe conditions in passageways/accommodations;

(6) clear evidence that watch keeping and other duty personnel for the first watch or subsequent relieving watches are impaired by fatigue.

<u>Clear grounds for a more detailed inspection</u> are, amongst others:

a. a report or notification by another authority;

b. a report or complaint by the master, a crew member, or any person or organizations with a legitimate interest in the safe operation of the ship, unless this complaint is clearly deemed to be unfounded;

c. other indications of serious deficiencies.

In addition there are requirements for expanded inspections for bulk carriers and tankers above a certain age and for passenger ships.

If the ship is found to comply, the PSCO will issue a 'clean' inspection report (Form A) to the master of the ship.

For the purpose of control on compliance with on board operational requirements, examples of "clear grounds" are, amongst others, the following:

a. evidence of operational shortcomings revealed during PSC procedures in accordance with (iaw) SOLAS, MARPOL or STCW;
b. evidence of cargo operations or other procedures not being conducted safely or as per requirement of IMO guidelines;
c. involvement of the ship in incidents due to failure to comply with operational requirements;
d. evidence, from the witnessing of fire or abandon ship drills, that the crew are not familiar with essential procedures;
e. absence of an up-to-date muster list;
f. indications that key crew members may not be able to communicate with each other or with other persons on board.

Furthermore, control on compliance with on board operational requirements may be included in the control procedures, particularly if the PSCO has reason to believe that the crew demonstrates insufficient proficiency in that area.

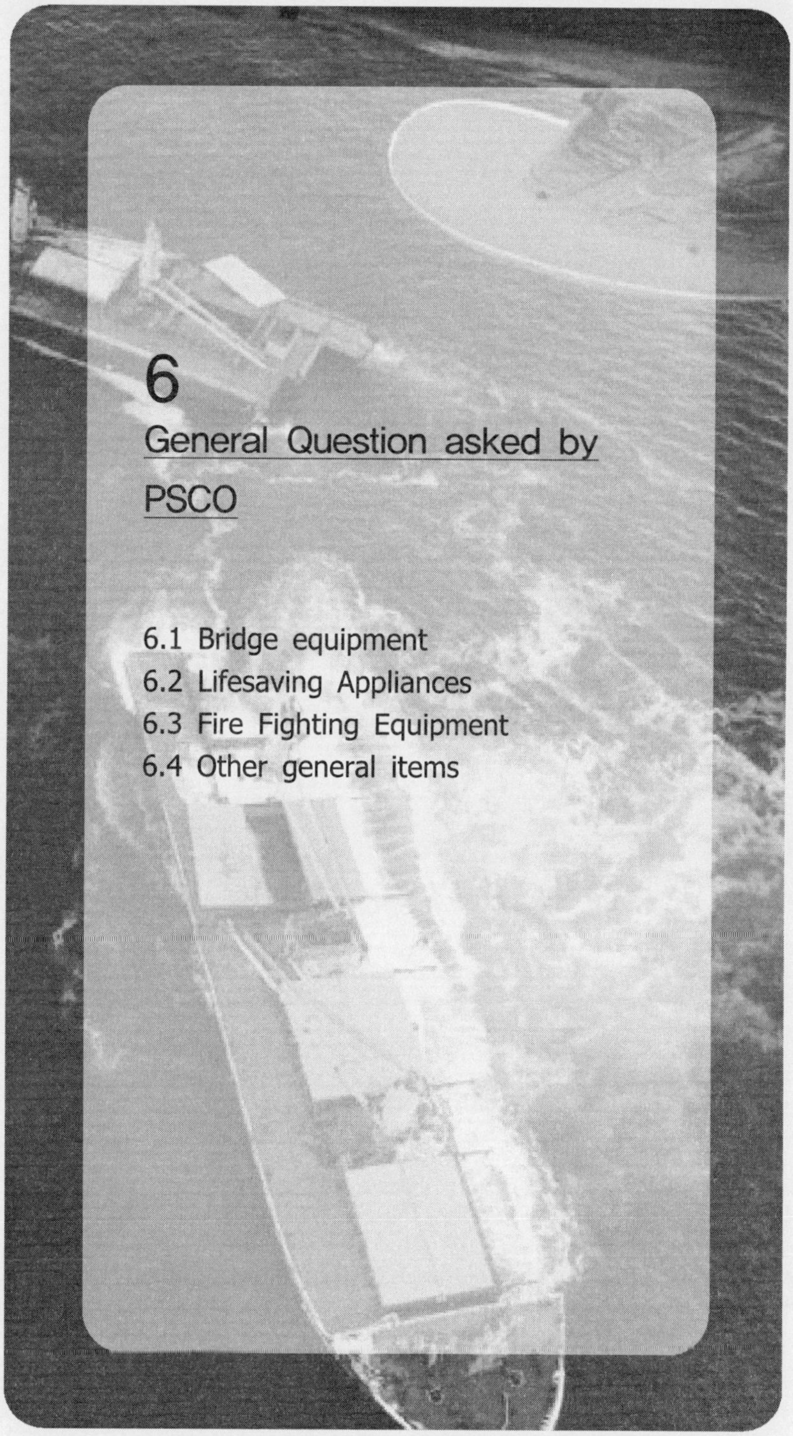

6 General Question asked by PSCO

6 - General Question asked by PSCO

PSCO will carry out a walk around after check ship and crew certificates in order to be sure everything in order with safety aspect and ship has not pose any danger to environment. PSCO will not examine every items detailed unless as explained previous chapters but will ask some simple question or will ask demonstration of working for important devices in order to be sure equipment are working properly and are in good condition. If will have any suspicious thing than PSCO will focus on that point. During walk around PSCO will ask questions and want to test some equipment. Down below tried to be given some examples of these questions and test, not all them but most important items given which likely PSCO may ask during walk around.

Before start question section need to be explained about previous action of PSCO. Before embark to vessel PSCO will look around the vessel and get a general idea about ship condition. At that time PSCO pay attention to freeboard marks and draft marks are they clearly marked and legible on shell plating or not. Some time freeboard marks been changed. In such case certificate should be proper and freeboard marks should be according load line certificate. And also PSCO will pay attention to general condition of vessel to get general idea about ship condition. After control ship' certificate upon the general idea during the embarking vessel, will carry out walk around and inspection.

Watchman(at gangway): Welcome on board. How can I help you?

Port State Control Officer"We are/I am …. PSC Officer. I want to talk with Captain or person in charge on ship now."

Watchman " May I see your ID card please?

PSCO "Off course"

PSCO's want to hear this question. Some times if there is 2 PSCO they change their ID cards between each other to be check watchman care and awareness. In such case they want to be sure that ship imply with ISPS code. PSCO should have an ID card how has holder picture and demonstrate holder authorization to carry out PSC inspection as per IMO resolution A.787(19) as amended.

Watchman: "may I have your name to fill up visitor log book. I will give you our ship's visitor ID,"

Some times writing the name could be hard in such case can be request to visitor write himself/herself name but need to be sure write correctly as per ID which submitted. Given to visitor ship visitor ID must not be forgotten and in that ID or separately with another card responsibility of visitor in case of emergency should be stated.

PSCO: Have a nice day Captain. I am … PSC Officer. I want to inspect your vessel.

PSCO before start inspection should introduce himself and should give information about inspection to ship master or responsible person/officer.

Captain"Welcome on board. Please have a sit. Would you like to drink some soft drink or coffee?"

PSCO: Thank you captain We can start directly to inspection. First of all we need to see ship's certificates and crew certificates. log books, LSA and FSS manuals, familiarization manuals, working hours….etc"

Ship certificates prove that ship is complying with international requirements. Incorrect or fake certificates subject to clear ground. PSCO control ships building date, size and international requirements upon to certificates. for example Safety equipment certificate gives information about ship safety equipment, this equipment's quantities and compulsory international rules upon year of ship construction. PSCO note this information and observe if this informations comply with actual while carry out visual observation (walk around) around ship.

Master: Of course here ship's certificate and crew certificates,. 3 mate will bring others log book...., etc now"

When PSCO control the ships logbook, he/she will look for below given logbook entries. The below given test and drills should be recorded properly into log book.

Steering gear testing and drills: Within 12 hours before departure. Emergency steering drills to be carried out every 3 months

Abandon ship drills and fire drills :Each member of the crew every month. Within 24 hours after departure if more than 25% of the crew have not participated in drills in the previous month. Each lifeboat to be launched and operated every 3 months. Rescue boats to be launched and operated each month as far as possible, but at least once every 3 months.

Down below given records can be fill up together drill records:

On-board training and instructions: Within 2 weeks after a crew member joins the ship, on-board training in the use of the ship"s lifesaving appliances including survival craft equipment. Instructions in the use of the ship"s life-saving appliances and in survival at sea to be given at the same interval as the drills.

Individual instruction in the use of the ship"s life-saving equipment and appliances to be given within a period of 2 months

Weekly inspections : Visual inspection of all survival craft, rescue boats and launching Appliances. All engines in lifeboats and rescue boats to be run ahead and astern for 3 min. Testing of the general emergency alarm.

Monthly inspection : Inspection of life-saving appliances and lifeboat equipment to be carried out using the check list required

Beside of this entries PSCO may check working language of ship which should be stated at the first page of logbook. Also the conning changing between officers and captain of ship should be stated in the logbook or should be recorded separately in port approaches, departures and in difficult waterways how presence of captain is compulsory and conning change required between OOW and captain of ship.

LSA and FFE Training Manuals: Provided in each crew mess room and recreation room, or in each crew cabin, complying with requirements. Language of manuels should be understood by crew.

Fire Safety Operational booklet: Provided in each crew mess room and recreation room, or in each crew cabin, complying with requirements and should be written in the working language of Maintenance plan should be kept onboard and should be complying with requirements

PSCO" Ok Captain there is not any problem with your certificates and log books. Now I want to walk around the vessel. Who will assist to me?"

Captain" 3 mate will assist to you and already informed engine room, responsible persons are waiting for you."

PSCO" Ok Captain thank you"

PSCO" First let's start from bridge 3/O."

3/O'" Yes, sure by this way"

6.1 Bridge equipment

PSCO: 3. Mate can you tell me who is in charge of safety and navigation?

This question ask to understand who is in charge of preparation of chart and other nautical publication's corrections, navigation equipment and passage plan preparation on board and to understand who is charge of maintenance of safety equipment. This could be changed upon to company procedure according to ISM so necessary responsibility distribution should be done. For example In some vessel chief officer could be safety officer but 3rd mate could be in charge of maintenance of safety equipment and 2nd mate could be in charge of navigation. Or chief officer could be safety officer but 2nd officer could be in charge of maintenance of safety equipment and 3rd officer in charge of navigation. But as usually 2nd mate navigation officer and 3 rd mate safety officer. This question important while carrying out make around by PSCO, there could be necessity of responsible officer during inspection.

PSCO: Can you show me nautical publications on board. (PSCO will control if compulsory publication are on board or not. Also will check correction up to date or not)

3/O: Here our ship publication all of them corrected and up dated

PSCO: How do you correct navigation charts? Can you show me chart correction log book and latest chart corrections?

This question ask to evaluate knowledge of officer if the answer will be proper than no need to detailed inspection on this topic. But if officer can not give proper answer than PSCO may focus on this point. So upon to case may need to show how to make chart correction and need to show log of chart corrections. In most of the case PSCO check latest correction in own port area, in important positions if the correction exist. Because of their own port area they know exact location and time of latest up dates.

PSCO: Can you show me voyage plan (it should be berth to berth)?

(Information written in the passage plan should be executed as practicable as, for example position fixing interval stated in passage plan has to be done if possible must be seen on the chart)

Passage plan should be signed by all the navigation officer and approved by master of ship. The information which entered to sailing plan should be correct and passage plan should be according company ISM manuels. If there is conflict between passage plan and manuels or incorrect information it may cause to clear ground.

PSCO: Can you show me GMDSS log book. Do you carry out necessary tests? Can you show me EGC message and navtex messages. Do you insert EGC and navtex messages on chart?

GMDSS equipment test entries to log book should be done properly. PSCO may want to approve correctness of entries. Because of this reason proper file keeping is important. Especially EGC files and VHF, HF/MF tests. Print out of test messages should be keep in other file or should be attached to gmdss log book properly. On the other hand for EGC message and Navtex messages necessary action should be taken. For example a navigational warning messages received during navigation and need to make correction on the chart, necessary correction should be done on the chart and than chart number which correction inserted, date of correction and sign of officer who make correction should be entered on the print out of EGC messages and must be filed. Navtex messages should be done by same way.

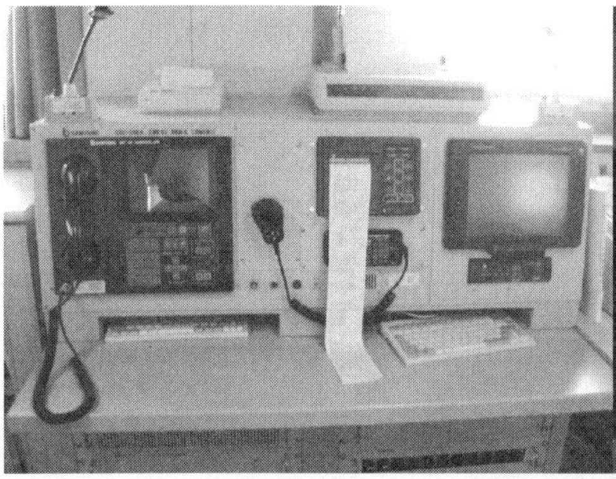

PSCO: Is the communication system working between W/H and E/R and between W/H and S/G?

PSCO may ask to carry out test of Communication system between Wheel House and Engine Room, W/H and steering gear room, W/H and Radio room. If there is 2 PSCO in this case they may go different location and may ask to ship crew to carry out test. Two-way communications which between emergency control station, muster and embarkation stations and strategic positions are not in common use and may get out of order or may not be in good condition.

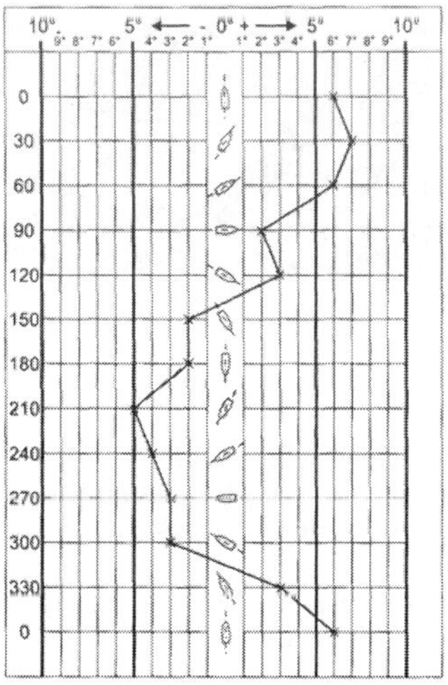

PSCO: Can you show me deviation table?

Deviation table should be prepared by authorized person. Authorized person might be captain of the ship and It should be carry out every year

PSCO: Can you show me repeaters and magnetic compass?

PSCO may check Magnetic compass and related devices. During Magnetic compass inspection PSCO will observe if Magnetic compass is clearly readable by the helmsman at the main steering position, communication between the standard compass position and the main steering position can be carry out properly, are there any Bubbles in the compass. Upon the importance of deficiency if found, ship can detain due to importance for navigation.

PSCO may check Gyro compass as well. PSCO will control if gyro compass can clearly readable by the helmsman at the main steering position or not, condition of the master gyro, and gyro repeaters for bearing with bearing device, master gyro and repeaters are synchronize or not. (Most of the vessel has not synchronize the repeater in steering gear room) PSCO may require the check gyro error log book as well.

PSCO: Is auto pilot, rudder indicator and rate of turn are in working condition?

PSCO may control Heading Control System (AutoPilot) if it is working satisfactorily or not. PSCO will check versatile change-over between manual and automatic and also NFU working or not. Change over procedure to emergency steering should be posted nearby Auto Pilot. PSCO may ask question on this point to OOW.

Also PSCO may control working condition of Indicators for rudder angle, Propeller RPM, thrusters' RPM and direction indicator and if available on board rate of turn indicator, are they working satisfactorily or not at the bridge and winks.

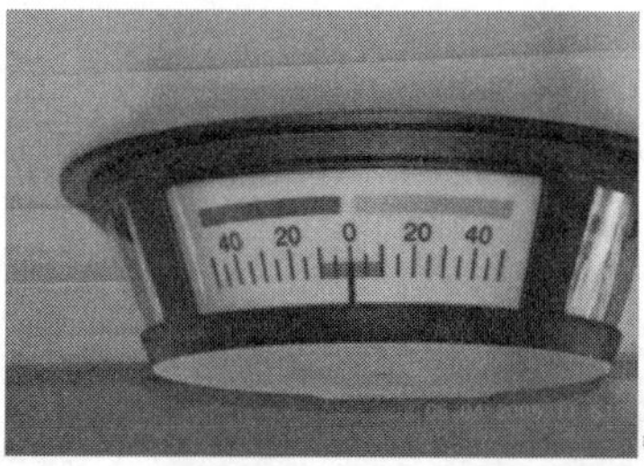

PSCO: Can you show me ship's GPS and AIS?

PSCO may check GPS and AIS if are they working properly or not PSCO may check data's for AIS if entered correctly or not as well.

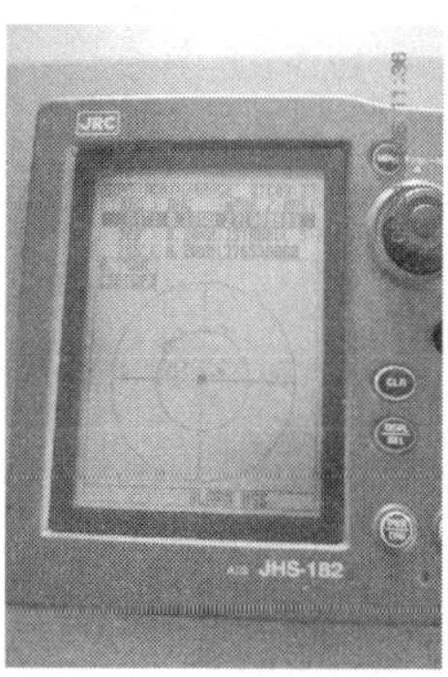

PSCO: Can you switch on radar?

PSCO will check Radar and ARPA which ever exist on ship. PSCO will observe are they working satisfactorily or not, are Plotting facilities available for Radar or not. PSCO may ask some question to OOW to understand knowledge of OOW if it is sufficient or not. For example, PSCO may ask about S band and X band radar. What is the difference between to band.

If the ship is equipped with ECDIS (If provided instead of navigational charts) PSCO may check Electronic Navigation Charts, Back-up System and Nautical data base. PSCO may ask some questions about how to operate ECDIS to Navigation Officers to evaluate ships officer knowledge if satisfactory or not for safe navigation.

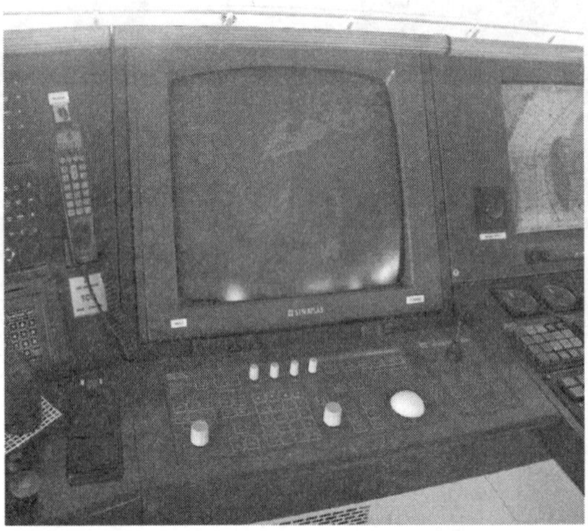

PSCO: Can you show me how do you test SART?

Radar transponders are one of the mainly controlled item. At least one radar transponder on each side should be provided and batteries should be valid. PSCO may ask to ship officer to show how to test SART. There is two point during the test: first, ships have X band and S band radar, if ship' officer does not pay attention to Radar band during test of SART, even SART is working properly, ship officer may not show and proof it. SART work on 9 GHZ which means X band radar can show SART signals, second is testing place, If the ship officer does not go to wink during the test of SART, PSCO may think that ship officer does not know how to test SART. Because during the test need to be at least 3 m away from radar. Most of the time PSCO ask such a basic test or question to evaluate ship officer knowledge if ship officer act accordingly or response properly in this case PSCO most probably will not ask any more question but if ship officer can not give correct answer or can not act properly in this case PSCO may dispute on that topic and can make further investigation.

PSCO: Can you show me how do you test EPIRB?

EPIRB must be test monthly accordingly instruction manuel, this test depend on manufacturer of EPIRB. Each EPIRB has different internal test procedure. PSCO want to know if the ship officer know how to carry out this test. Recently most of the ships officer has GOC certificate so the officer who has GOC certificate should know this procedures. This test should be carried out according to GMDSS and some time PSCO may ask the procedure for canceling of EPIRB distress signal which is send by mistake during the test of EPIRB. If a ship send a distress signal from EPIRB by mistake and did not carry out necessary procedure for false emission cancelation, subject to Port state control and penalty. Also annually print test should be carry out and certificate or print out of test should be kept on board. EPIRB should be installed to float free area (must not be restricted to float) if the ship sink should be able float up to sea surface.

PSCO: Can you show me emergency VHF's and spare batteries?

PSCO may want to see two-way VHF radio-telephone apparatus on bridge. At least three sets complying with the standards should be provided with spare batteries and operation of the apparatus can be controled. In some cases has been observed that two-way VHF (emergency VHF) are in use of common purpose which is not allowed and also spare batteries have been used. Spare batteries must be sealed and expire date should be valid.

PSCO: Can you show me bridge pyrotechnics?

PSCO may check existence of distress flares. At least 12 parachute rocket signals should be available and valid on board. Also Line-throwing appliances may be controled by PSCO. On board should be provided four rockets capable of carrying line at least 230m. Proper instruction should be available and validity of rockets subject to control. PSCO may ask to ship officer that how to operate line-throwing appliances. Ships crew should be familiar with operation of ship' equipments.

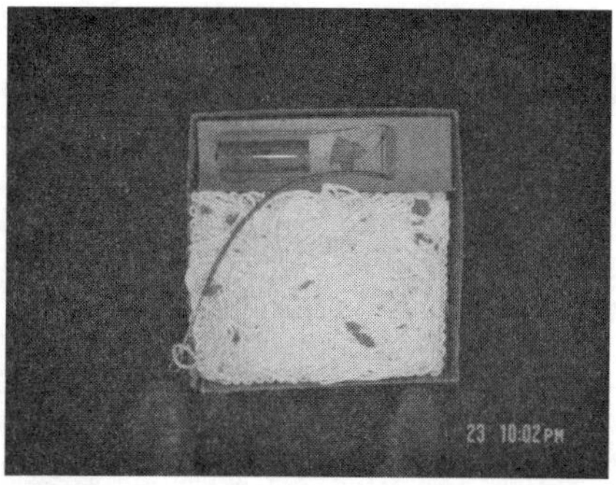

PSCO: Are doppler log and echo sounder in working condition?

PSCO may control Echo sounder, Speed and distance log through water, Speed and distance log over ground, are they working satisfactorily or not. Here is one important point for PSCO to understand knowledge of ship officer. In the ARPA, speed should be obtained from log through water. The reason is calculation of CPA and TCPA. If the speed obtained from GPS or log over ground cause of current effect CPA and TCPA can be calculated wrongly. This can cause the accident. PSCO may ask how do you obtain speed for ARPA. Most of the ARPA has a function to choose speed log.

Before switch on and switch off Echo sounder, date, time and location should be entered on the paper. In case of accident it may be necessity.

PSCO may check this point as well.

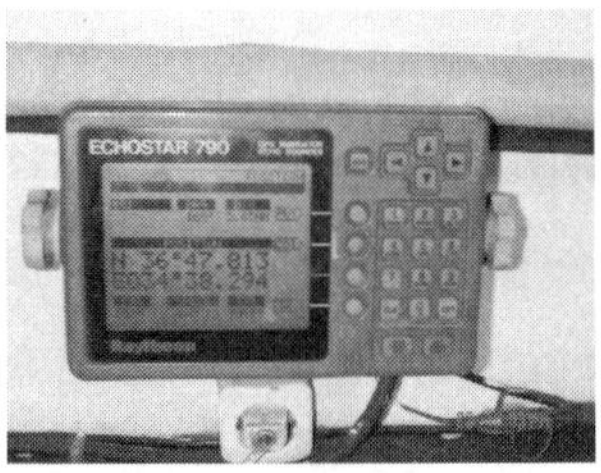

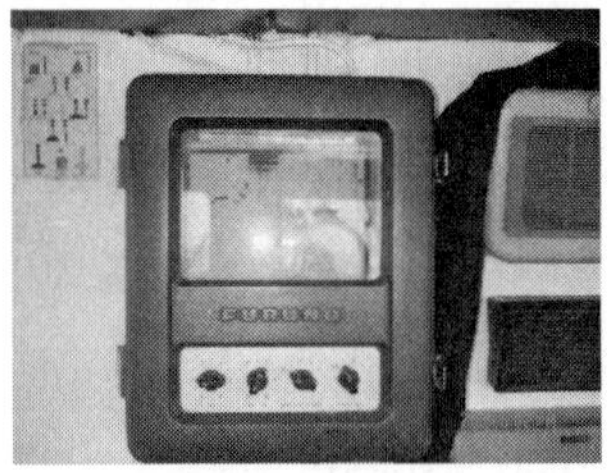
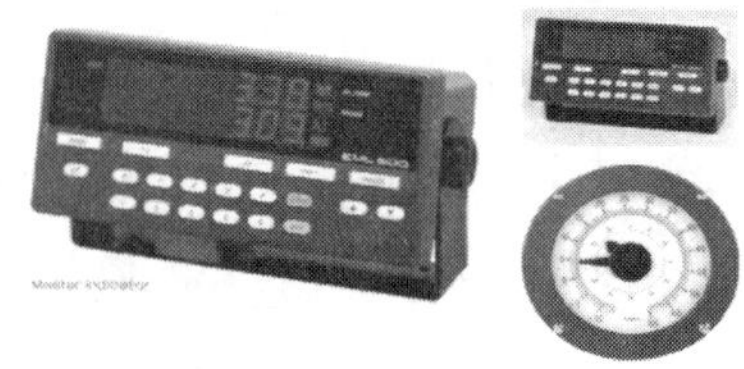

PSCO: Can you show me Daylight signal?

PSCO may control Daylight signal(ALDIS) if it is working satisfactorily or not, if it is supplied from emergency power or not and PSCO may check spare bulbs are available or not. ALDIS should be positioned in a position how can be use easily.

PSCO: Are the navigation lights in working condition together with emergency power supply?

PSCO may check navigation lights. (Fore & aft masthead lights, Side lights, Stern light, Anchor light, Not under command light) and distribution panel for navigation lights. Also PSCO may control do navigation lights has spares and are they working with emergency supply.(24 volt).

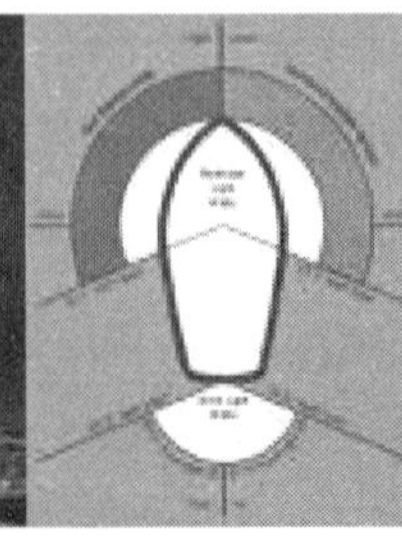

PSCO: Do you have bell, gong and black ball shapes?

PSCO may check forecastle bell and gong (ships of 100m and upwards in length) are they available on board and working satisfactorily. Also PSCO may check for black ball shapes at least 2 sets should be available on board. Most of the case have shown that ship officers do not know location of this equipment or due to negligence equipment found not in good condition.

PSCO: Can you tell me where is ship's maneuvering poster?

Maneuvering poster of the ship should be posted at the bridge.

6.2 Lifesaving Appliances

PSCO: 3 mate let see life boat and its equipment

PSCO most probably will control Lifeboat and/or rescue boat during walk around. PSCO will control condition of hull inside & outside (no rust, no doublers, no hold, no damage,..), grab lines on both side in order or not, bilge keel rails on both side in good condition or not, rudder stock, rudder and tiller and stern frame in order or not, thwarts, side benches, clutch holes, gunwales are in good condition or not, engine, foundation, exhaust pipe, propeller and shafting with clutch are working or not (PSCO may ask to run engine ahead/astern and test rudder to understand condition and if has any dispute or for be sure may ask to carry out a drill), reflective tape on hull, marking (Ship''s name, No of persons,

Registry of port etc.), retro-reflective tapes, plug with packing and a chain with indication of position, bilge pump with hose (testing) and lifeboat and/or rescue boat equipment are in good condition or not. (Briefly there must not be any damage at L/B. For example enclosed type L/B window has crack PSCO may detain vessel or may be require to be repaired before departure. Another example can be given about L/B equipment. Magnetic compass which in the L/B can have bulb inside if it is water type, most of the case have shown size of the bulb is too big in such case it can be assume as deficiency. Safety Officer of the ship should check all the equipment regularly if there is any missing, damaged or improper item)

PSCO will check also launching arrangement of life boat and rescue boat condition. During this control PSCO will pay attention to condition of davits, condition of blocks, falls, padeyes, links, fastening and all other fittings (Falls to be turned end for end at intervals of not more than 30 months and fall to be renewed at intervals not exceeding 5 years), condition of brake (Winch), condition

of release gear, condition of skates and fenders, condition of embarkation ladders, handholds, side ropes, steps and fitting shackles/padeyes and condition of boat lights by emergency power.(Brake (Winch) to be thorough examined at intervals not exceeding 5 years and on-load release gear to be overhauled and tested under a load at intervals not exceeding 5 years and should be certificated, if PSCO has dispute can require to see certificate or during control of ship's certificates might be checked already)

PSCO: Can you show me life rafts?

Inflatable liferafts are another important items. Liferafts to be serviced at intervals not exceeding 12 months with release gear except when extension is authorized by the flag state. Container of liferafts to be marked with; maker''s name, serial No., last service date, No. of persons, etc. PSCO will control this item visually and there must not be any damage/hole.. etc at the container of liferafts. Also stowage of liferafts will be controled. Condition of stowage, proper fitting of weak link (in case of a type without weak link, notice to be posted) condition of embarkation ladder are important points. Also HRU and launching device of liferafts should be in good condition. Some ships have HRU and connection of HRU not correct. Because of this reason PSCO most probably will check HRU connection.

PSCO: How many lifebouys do you have on board?

Number of lifebuoys can change upon to length of ship. It can be summarize as ships of L(m) and No. of buoys as given

less than 100m ship required 8 pcs

less than 150m ship required 10 pcs

less than 200m ship required12 pcs

200m or over ship required 14 PCs

At least half of the total number of buoys to be provided with self-igniting lights (SIL) and at least one buoy with a buoyant line on each side should be rigged (30m or twice the height at stowage position above water level, whichever is the greater) Proper No. of lifebuoys with marking and retro-reflective tapes should be available on board. PSCO can observe these rules if ship is comply or not. On the other hand PSCO may check is there two buoys with self-igniting lights and smoke signals being capable of released by quick release gear, having a mass of at least 4kg on bridge wings (MOB). Illumination of MOB, validity of smoke signals, operation of release gear can be inspected.

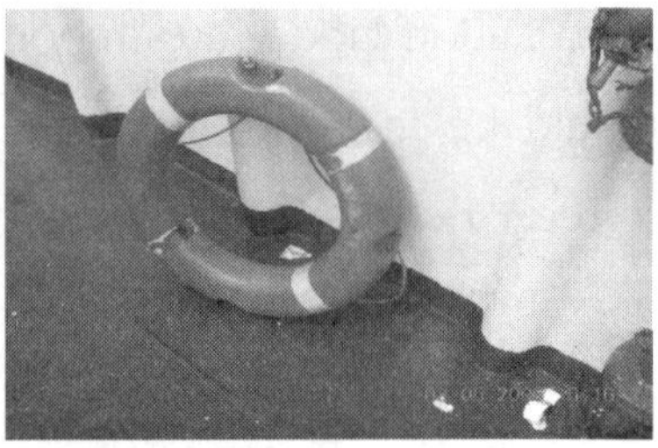

PSCO: How many lifejackets do you have on board?

Lifejackets should be available for every person on board with retro-reflective tapes, a whistle & light and additional lifejackets should be ready for persons on watch and for use at survival craft stations. PSCO may visually observe/check whistle, light and additional lifejackets randomly.

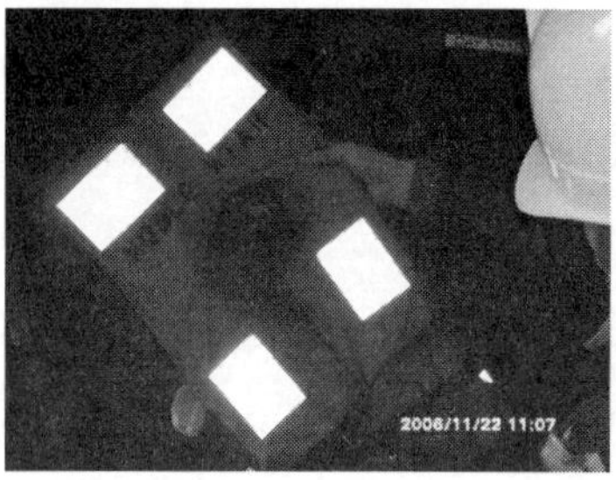

PSCO: How many immersion suit do you have on board?

At least 3 sets of immersion suits should be avaliable for each lifeboat on board the ship, and should be ready for immediate use. PSCO may check immersion suits light, whistle, and condition if are they in good condition or not. Some times immersion suits are not ready to use, for example zip of the suit could be stuck or seized. If immersion suits do not provided for each person on board, thermal protective aids should be provided to each person on board and must be ready for immediate use. PSCO may ask to see TPA to confirm that it is provided to each person on board.

6.3 Fire Fighting Equipment

PSCO: How is condition of fire doors? Are they in good condition? Do you maintain properly?

Fire doors are important tools for fire fighting. PSCO may check operation of doors if satisfactorily or not. Self-closing doors should be operable and must not to be fitted with hold-back hook

PSCO: How many fire dampers do you have? How frequently do you control them?

Fire dampers are another important tools for fire fighting. PSCO may check Operation of dampers if satisfactorily or not. in cargo holds, engine room, accommodation spaces, control stations and other spaces. Clear marking of “"Close-Open”" should be marked. Also general condition of damper should be in condition.

PSCO: How many fire pumps do you have on board? Can you rig 2 fire hose and can we test emergency fire pump?

PSCO may check main fire pumps, if it operate satisfactorily or not. Proper pressure should be maintained and pressure gauges should be in good order.

Emergency fire pump is one of the important item which can be check during PSC inspection. PSCO may want to see operation of pump if satisfactorily or not. During operation proper pressure should be maintained, pressure gauges should be in good order and instruction manuel should be posted in vicinity of the pump. Ships crew should be familiar with starting and running emergency fire pump. As mentioned before if PSCO want to see operation of emergency fire or fire pump, may request to prepare/rig randomly chosen two fire hose and than run the pump. It could be ask to demonstrate during fire drill and for emergency fire pump testing may be request black down to ship in order to be sure emergency fire pump not supported/feed by fire pump. During the test of fire pump or emergency fire pump fire also main piping may be control by PSCO. During control of fire piping there must not be observed any leakage, heavy wastage in lines or doublers, clamps, soft patches in fire pipe line. At the same time PSCO may check fire hydrants. Actually PSCO does not check one by one all the fire items, during walk around the vessel. PSCO look around and observe general condition of this item. PSCO may look if fire hoses easily coupled to hydrants, fire hydrants valves satisfactorily operable or not and valve handles are broken or not, rubber gaskets condition are good or not. On the other hand PSCO may check general condition of stowage boxes of fire hoses and nozzles, if stowed in good condition and easily usable or not, clearly painted (red color) and marked or not. Unless inspection detailed PSCO will not check all the fire hose, hydrants, nozzles and other items one by one. PSCO will check randomly some fire hose if complete set with nozzle and couplings, hoses in good condition, without leakage. If PSCO find some hose not in order may detail inspection. Also PSCO will check fire nozzles if in good condition, without leakage. Fire nozzles should be Jet type nozzles, and jet/spray dual type in engine room. and Jet/spray dual type with shut-off device for all nozzles. All the nozzles should be able easily change mode and shut-off device.

PSCO: How many fire extinguisher do you have on board?

PSCO may check portable fire extinguishers (foam, dry power, CO_2 ....) as well. During control PSCO may check the number of portable fire extinguishers of each type according to the fire control plan. Also cylinders, hose and handle are in good condition, without serious corrosion/damage or not and extinguisher service date and cylinders pressure tests are valid or not. If has any dispute may ask to see certificates.

Foam type fire extinguisher of 135 litters capacity in firing space of boiler and in spaces of fuel oil system and foam type fire extinguishers of 45 litters capacity in engine room should be available. PSCO may check visual conditions are in good order, without wastage and is it easily usable condition or not. Validity of pressure test and service dates may be observe.

PSCO: Do you have foam applicator on board?

If PSCO check the portable foam applicator unit, may check the air-foam nozzle, portable tank of foam making liquid, and one spare tank. PSCO may test the connection to fire main by a fire hose. Condition of stowage container should be in good order and foam test should be valid.

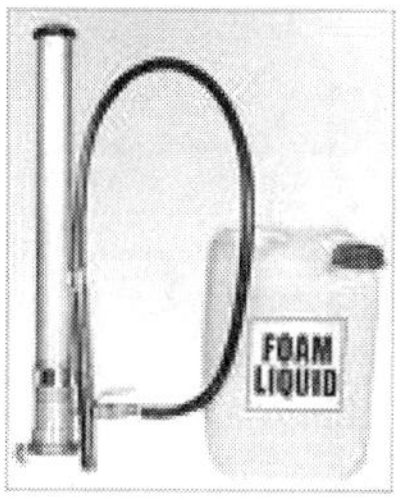

PSCO: What type of fixed fire extinguish system do you have on board?

Fixed fire extinguishing arrangement in E/R and cargo spaces (CO_2 or Halon, foam, water spray) are subject to control by PSCO. PSCO may check piping in lines in good order, without leakage or not heavy corroded. Regular check of lines by air blow or water flow test at Class Special Survey and Intermediate Survey and also CO_2 or Halon cylinders to be level/weight measure test during Class SS and IS should be done. PSCO may ask to see certificates of this tests. Validity of foam liquid is 5 years. After 5 years, effectiveness of foam liquid to be checked, and a sample test certificate should be available on board, PSCO may ask to see this certificates too. PSCO may request testing the audible alarm for the release of gas (CO_2 or Halon)

PSCO: What type of fire detectors do you have on board?

Fire detection alarm system is also subject to PSCO inspection. Regular checking of the detection system and fire alarm should be done on board. PSCO may test some detector or may want to see demonstration of a test by ship crew.

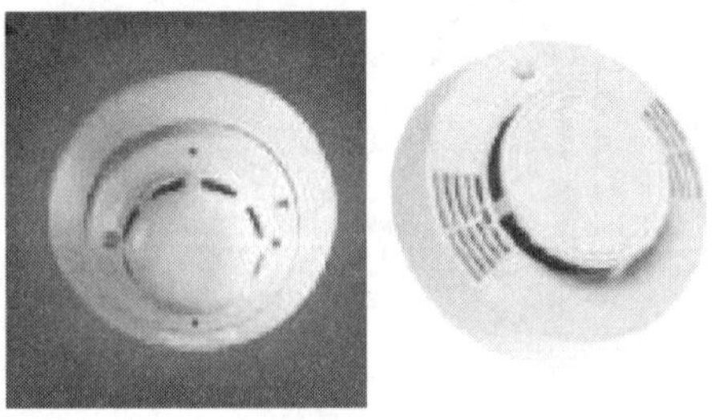

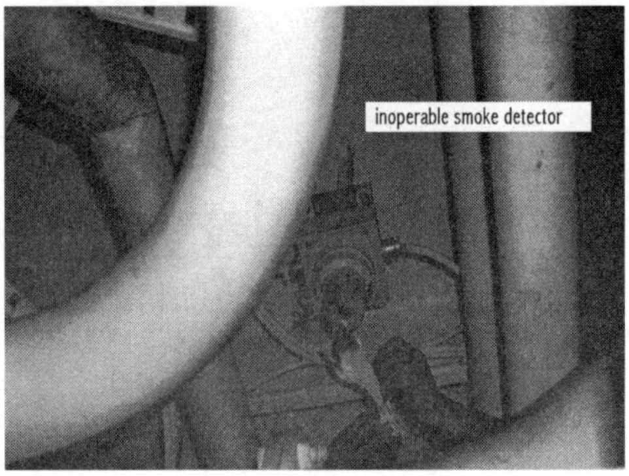

PSCO: Can you show me shut-off valves? Who carry out test of shut-off valves?

Fuel oil tank shut-off valves should be provided for over 500 grt ships built on and after 1 July 1995 and over 1000 grt ships built before 1 July 1995. All valves to be closed by remote control from outside of E/R. Demonstration of satisfactory operation of valves can be request by PSCO. Shut-off valves are operated by air, the air cylinder is always charged with correct pressure and pressure gauge should be in good condition which can be controled by PSCO. Test of shutt-off valves should be carry out regularly by engineers.

PSCO: Can you show me emergency stop of fans and fuel oil pumps?

Emergency stop of fans and fuel oil pumps in E/R should be available. PSCO may want to see operation of emergency stops if in order or not.

PSCO: How many Fireman's Out fit set do you have on board? Where are the locations of FO, can you show me?

PSCO may check Fireman''s outfit. Two(2) sets for cargo ships and four(4) sets for tanks should be available on board. Stowage conditions should be in good order according to the fire control plan. Protective clothing, boots and gloves, helmet, electric safety lamp, axe, a fireproof lifeline of sufficient length for each breathing apparatus with a snaphook, breathing apparatus with a smoke helmet or smoke mask and air pump, with proper length of air hose, or a self-contained breathing apparatus should be together with FO sets and also 200 % spare air cylinders should be available on board. PSCO may ask to put on FO to designated person who should wear FO in case of fire. Designated person should be able to wear FO in a short time, if designated person can not wear or takes long time, PSCO may request to carry out a fire drill. Ships crew should be familiar with their duty and ship's equipment for safe operation.

PSCO: Do you have fire blanket?

Fire blanket should be provided for galley. Must be in good condition together with IMO sign. In some ships heavy dirty, worn, holed and useless blanket founded.

PSCO: Do you have international shore connection?

PSCO may check International shore connection, especially in USCG inspection coast guard inspector control ISH. At least one(1) shore connection with standard flange dimensions should be available on board. Together with flange, four sets of bolts and nuts, each of 16mm in diameter, 50 mm in length and one gasket packing should be available on board.

PSCO: Are the emergency light working? How do you support emergency lights? Do you have emergency generator or batteries for emergency lights?

Emergency lights should be provided with 24 volt. PSCO may control if emergency lights are working satisfactorily, PSCO may control lighting condition in engine room, accommodation spaces, wheel house, control stations, outside passage. Bulbs and glasses should be without damage.

PSCO may require to test of emergency generator if exist on ship. During these test PSCO will control Operational test and confirm of F.O. tank level. Condition of starting devices and ease of operation by crew should be provided. This test required to confirm condition of equipment and crew familiarity with equipment.

PSCO may check emergency batteries if exist on board. Battery room should be clean and record of batteries should be kept properly.

PSCO: How is the condition of emergency escapes?

Means of escapes should be ready for immediate use. Steps and handrails should be without damage and lighting should be operates satisfactorily. PSCO may control IMO signs if posted properly or not. It has been recorded some ships have been detained due to obstruction in fronts of means of escape.

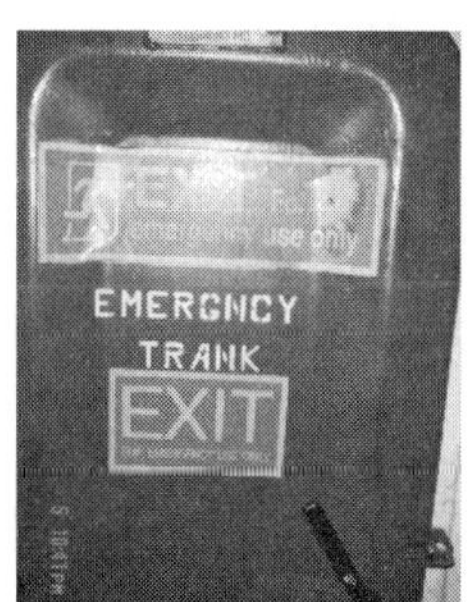

PSCO: How many EEBD set do you have?

Emergency Escape Breathing Devices (EEBD) should be provided as per fire control plan & flag state requirement. Number of EEBD can change upon the flag state. PSCO may ask to see or check EEBD sets. At least one should be spare for training purpose and should be in good and easily usable condition. Suitable maintenance should be maintain according to the Manufacturer' instruction and IMO signs should be posted.

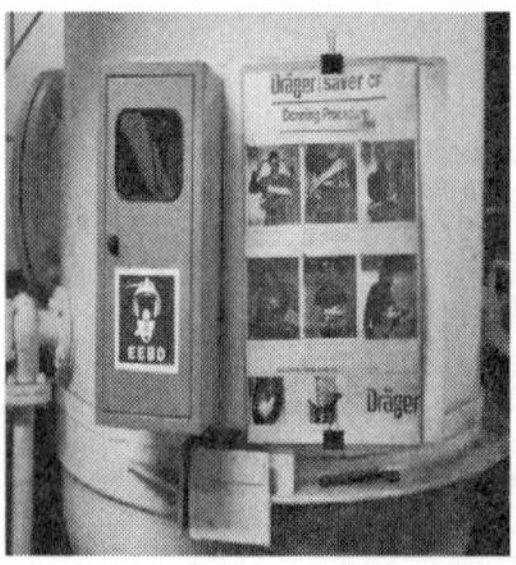

6.4 Other general items

PSCO: How is the condition of weathertight doors and closures?

While PSCO walking around most probably will pay attention to doors of all access openings in bulkhead at ends of enclosed superstructures. They must be effective weathertight, no heavy corrosion, holes and condition of gaskets and clamping devices should be in good order

Same as door and other structural condition during walk around PSCO may pay attention such as ventilators. Ventilators coamings should be in good condition without heavy corrosion, holes. Closing covers should be in good condition, efficiently weathertight and gaskets and clamping devices should be in order.

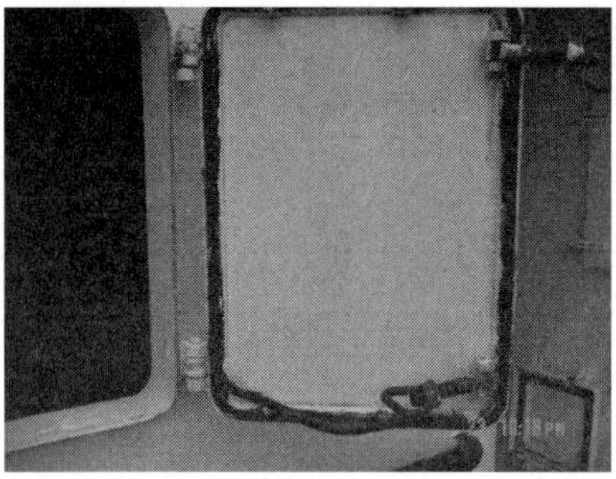 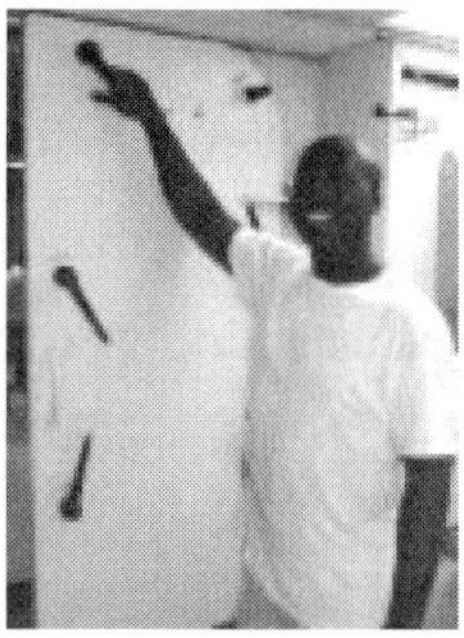 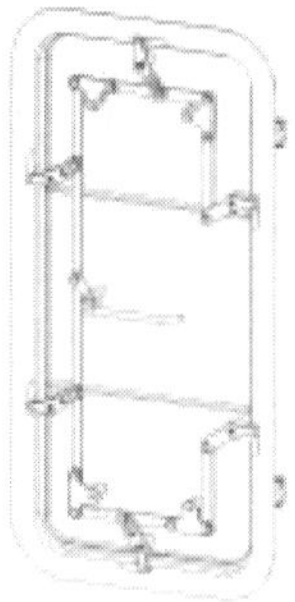

PSCO: let me see overflows conditions!

Air pipes (overflow) should have marking and coamings should be in good condition without heavy corrosion or holes. PSCO may check also air pipe floats and wire gauzes how should be in good order.

PSCO: Lets have look to mooring arrangement?

Mooring winches should be in good condition. Brake bands, foundations, grating plates should be in good condition, no abnormal wear, no wastage, missing or broken sections. Sufficient ropes should be available on board and should be in good condition and also capstans should be operate satisfactorily. PSCO may check all these items during walk around. On the other hand windlass's winches, brake bands, foundations, grating plates should be in good condition (no abnormal wear, no wastage and missing or broken sections)

PSCO: Do you fallow garbage management plan?

Garbage management implementation another item for PSC check during walk around. PSCO may check garbage management plan and log while checking ship's certificate but during walk around PSCO may check if placard has been stick and garbage separation and collection carry out according to garbage management plan on board.

PSCO: Do you carry out SOPEP/SMPEP drills? Can I see SOPEP/SMPEP equipment?

SOPEP/SMPEP drill and equipment can control by PSCO. SOPEP/SMPEP equipment may change ship to ship upon ship manuel. first PSCO will check manuel after that will control equipment. Detailed information written in the manuel. PSCO will check contact list if up to date or not (it can be update through internet every 3 months)and also PSCO will check tanks capacity and information entered correctly or not by compare ship's certificate. For the drill PSCO will control that; ship-related persons should be involved in the drill covering all parts of SOPEP/SMPEP which should be carried out at regular intervals. PSCO also may control records of drills.

PSC Officers mainly control other records and plan as well. Down below have given some samples :

Fire control plans: Permanently exhibited in accommodation spaces. Permanently stored in watertight cases outside the accommodation main entrances.(Port & Starboard) Language should be understood by crew. Fire control plans should be kept up-to-date

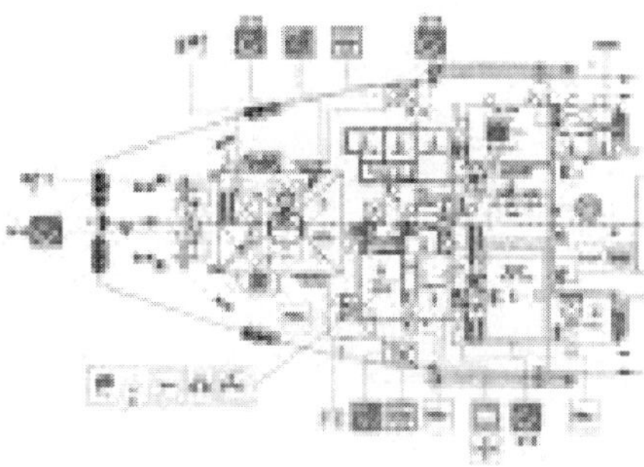

Muster List : Exhibited in W/H, E/R and crew accommodation spaces. To show duties according to Reg.III/37 Language should be understood by crew.

Posters or signs :Provided on or in the vicinity of lifeboats, liferafts, rescue boats and their launching controls. Symbols should be used according to IMO Res.A760(18)

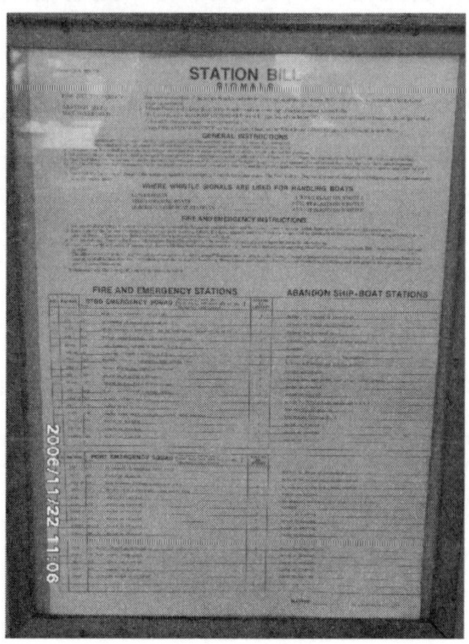

Pilot ladders: Condition of side ropes, rubber steps, wooden steps should be in good order and Proper handholds should be available.

Maintenance plan should be kept onboard, complying with requirements of Reg.II-2/14

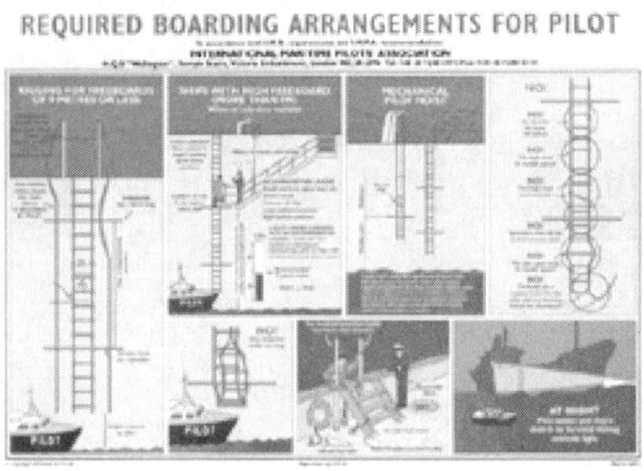

Fire pump test: PSCO may ask discharge test of fire fighting system by operation of main fire pumps or emergency fire pump separately. During this operational test of main fire pump or emergency fire pump separately PSCO will check sufficient delivery pressure which should reach distance 12m and over Isolation valves how should be operable. When pressure measured, required pressure is 0.27 N/mm^2 for ships 6000 GT and over 6000 GT and 0.25 N/mm^2 for ships under 6000 GT. During the test should be observed no leakage of fire lines, also confirmation of F.O. tank level for emergency fire pump engine might be carry out. Ease of operation by crew also will be observed by PSCO. In most of case PSCO require to rig to fire hose and observe fire line and hose. If PSCO has any dispute about emergency fire pump, may require to black down to generators in order to prevent feed emergency fire pump from fire pump. On the other hand this operation could be required to observe condition of equipment which must be feed by 24 volt in emergency cases. Such as navigational lights, emergency lights....

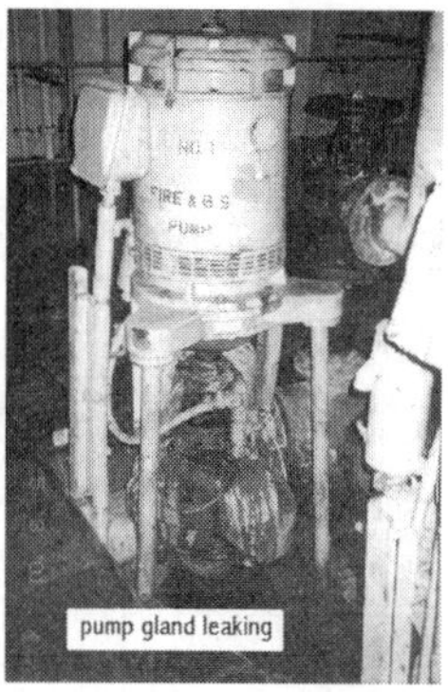

Steering Gear Drill: During walk around PSCO will control Steering gear (S/G) system. PSCO will control operation of main and Aux. S/G (full movement of the rudder), Remote control system, Emergency power supply, Rudder angle indicators either it is in relation to actual position or not, automatic isolating arrangement, Visual inspection of S/G and connecting linkage and operating instructions with block diagram in W/H, S/G room, also PSCO may require testing of alarms and operation of S/G manually by ship crew in order to see crew familarity and capability for emergency cases. Emergency steering gear drill should be carry out every 3 months during drill need to be practiced of emergency steering procedure (including direct control, communication, alternative power) and also within 12 hours before departure and arrival port tests and inspection should be carry out by ship crew. PSCO will inspect and ask questions due to these requirement.

Abondon Ship Drill: As mentioned before record of drill should be kept properly. PSCO may ask to carry out an abandon ship drill to be observe crew performance or in case of have some dispute with ship carry out regular drill. On the other hand, PSCO may require to carry out drills if he/she thinks FFE and LSA are not in good order. While carry out drill most probably deficiency will come up, cause of these reason PSCO may require to carry out related drills. Abandon ship drills should be carried out (every month and within 24 hours of departure if 25% of the crew have not participated on board the ship in the previous month)

During drill summoning of the crew to muster stations with emergency alarm according to the muster list will be observed. After that confirmation of the duties stated in the muster list will be observed. Also some other item such as worn of lifejackets correctly by crew, lowering of at least one of life boat, starting & operating the lifeboat/rescue boat engine(s), test of emergency lighting, running of propeller of lifeboat, general overview of lifeboat if there is any damage, bend, .…….etc exist might be observed by PSCO. Also record of launching and maneuvering by the assigned crew in the water every 3 months of life/rescue boat will be checked.

Same as Abandon ship drill, Fire drills should be carry out every month and within 24 hours of departure if 25% of the crew have not participated on board the ship. PSCO may require to carry out Fire drill in this case PSCO will check: Summoning of the crew to stations according to the muster list, Checking fireman"s outfits and other personal equipment, including fitting on crew member in turn, Checking the communication equipment, Checking the operation of fire door, watertight door, fire dampers and main inlets/outlets of ventilation system, Operating shut-off valves of F.O. tanks and emergency stop of fans and starting a main and emergency fire pump in turn, and discharging test using the two jets of water. PSCO may require to test or use emergency Fire pump. It has been mentioned in previously. If PSCO has dispute or want to be sure about Emergency Fire pump, PSCO may require to black down generator and run only emergency fire pump.

Appendix

A : Some sample deficiency photos

B : Sample Checklist

Appendix A : Some sample deficiency photos

- Fire construction:

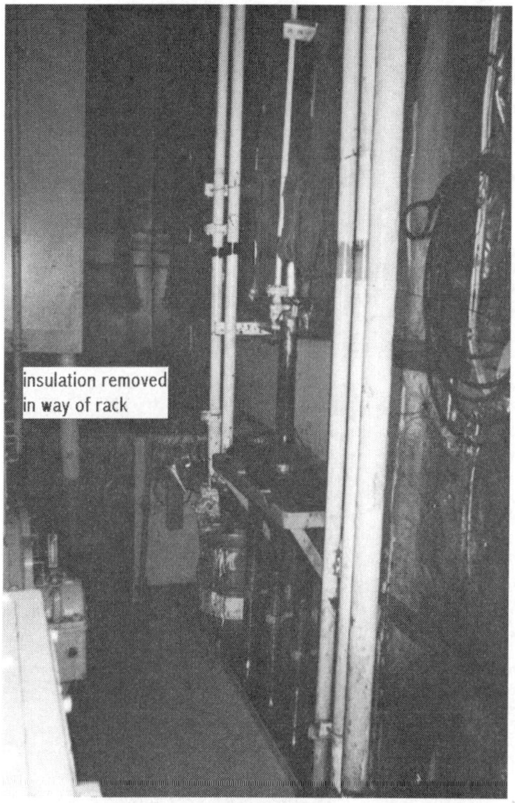

insulation removed in way of rack

hook for keeping self-closing door open

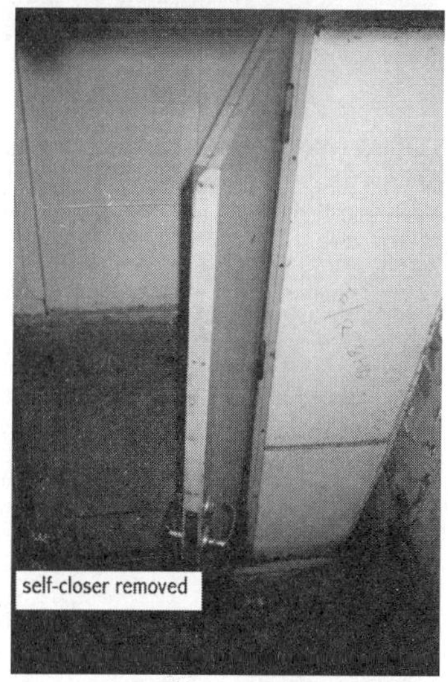

self-closer removed

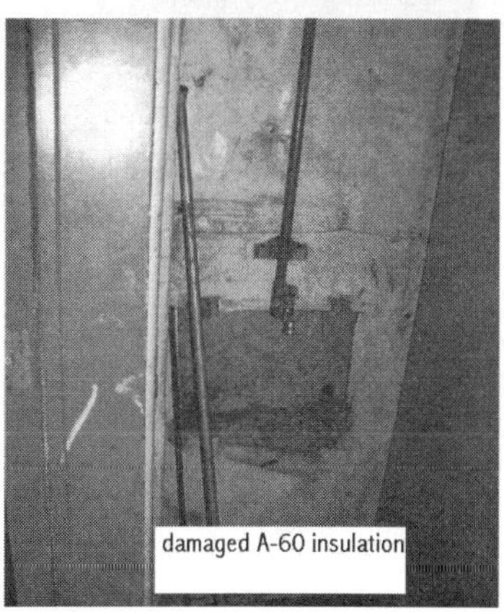

damaged A-60 insulation

defective service hatch between mess room and galley

- **Fire damper:**

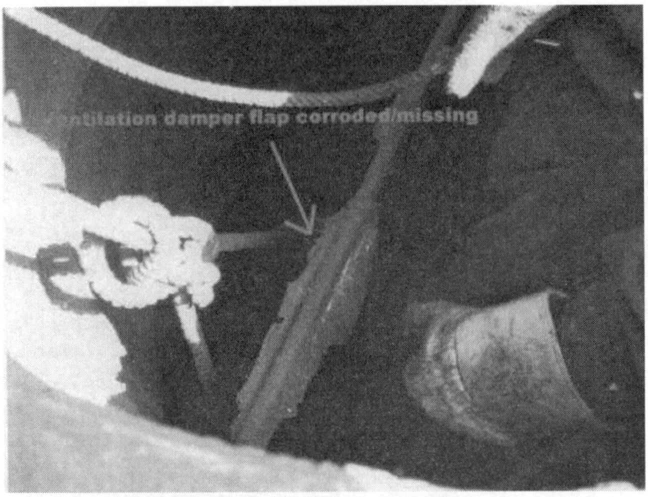
ntilation damper flap corroded/missing

hole
hoses leaking

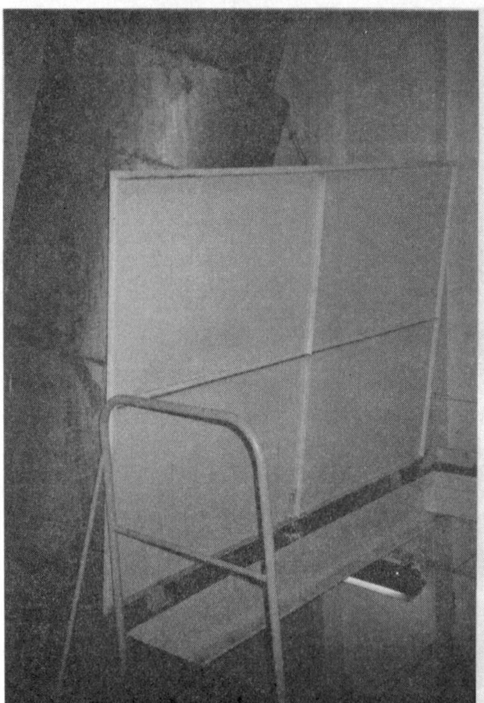

air sylinder for the damper at left missing

- **Fire Detector:**

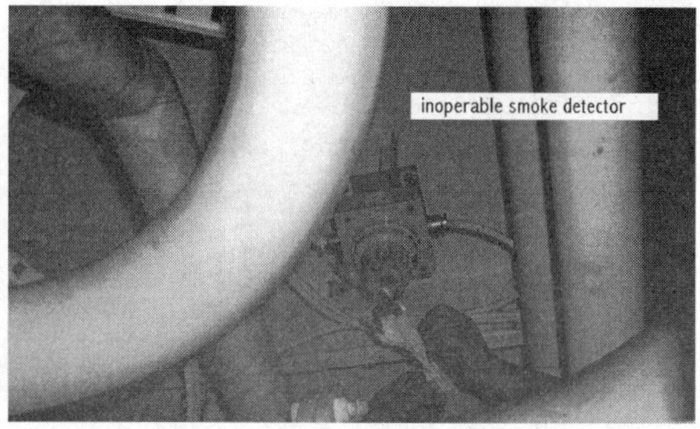

- **Fire Extinguisher:**

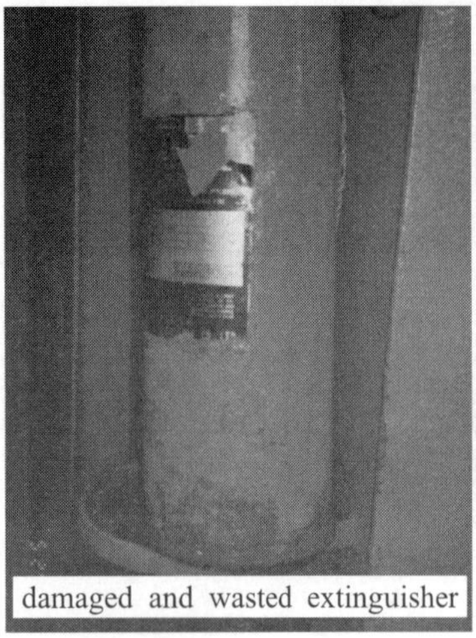

damaged and wasted extinguisher

fire main hydrant leaking

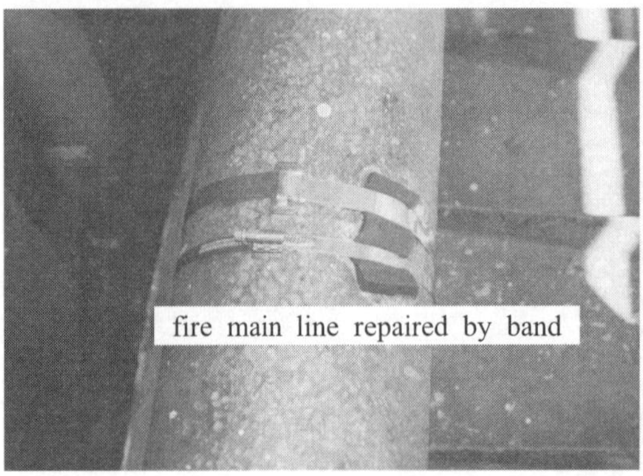

fire main line repaired by band

handle broken fire hydrant

- **Fire pump**

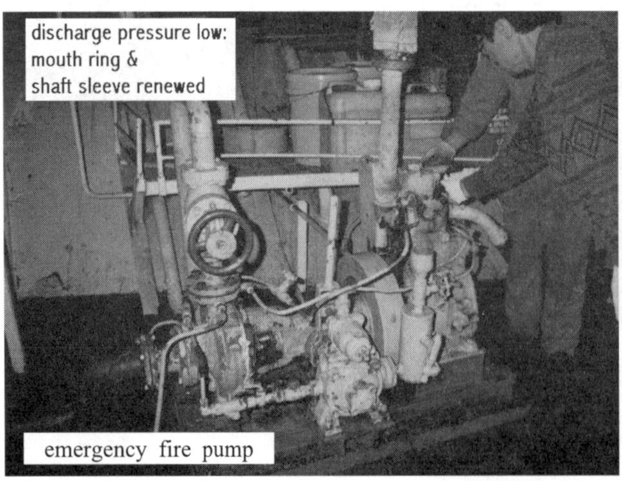

emergency fire pump

fire pump valve seized

pump gland leaking

leaking gland

- Fireman Outfit:

dirty fireman's suit

locked safety locker

- **Fire hose-nozzle**

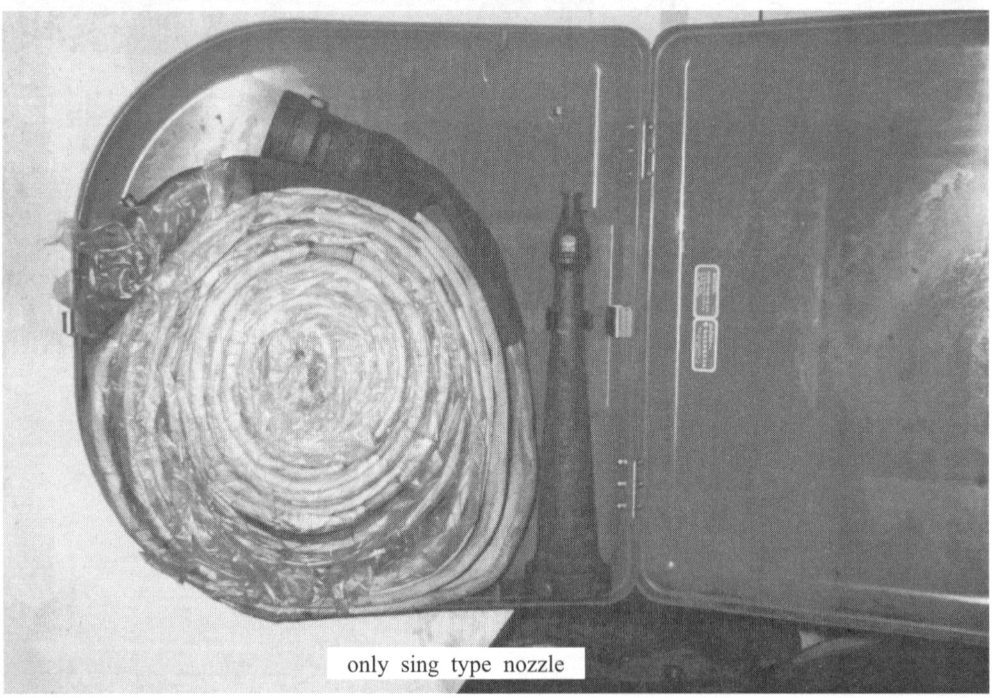

• **Embarkation**

seized light stand of lifeboat embarkation position

embarkation ladder not complying with SOLAS

wasted connection of
embarkation ladder

wasted frame

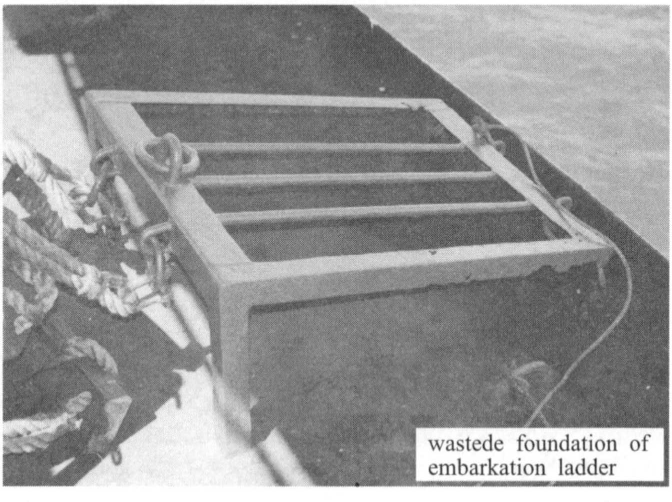
wastede foundation of
embarkation ladder

- **Equipment**

broken lens

EPIRB installed inside

expired hand flares

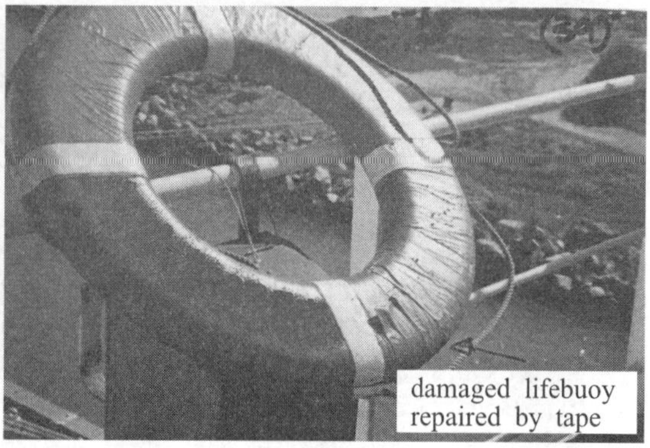
damaged lifebuoy repaired by tape

insufficient arrangement for quick releasing system

inappropriate arrangement of quick releasing system

- **Life Boat**

wasted aft hook of lifeboat

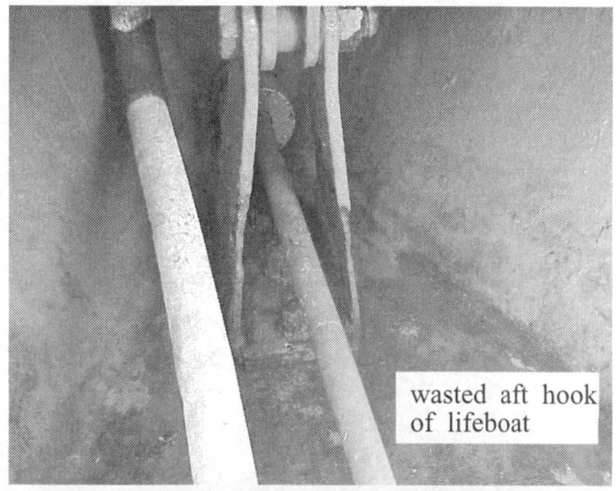

wasted aft hook of lifeboat

Broken bilge keel

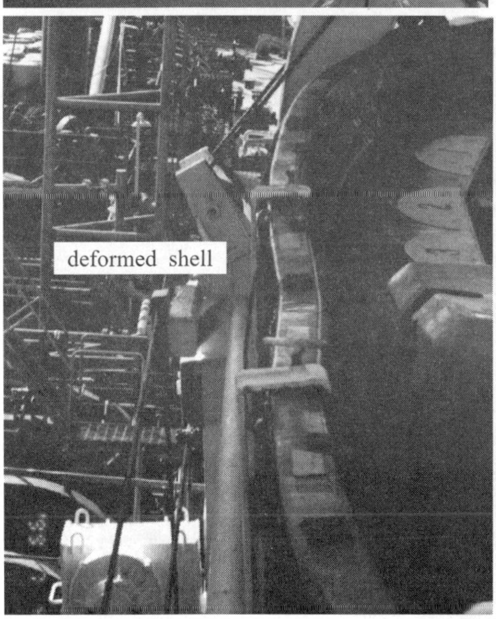

deformed shell

lifeboat engine inoperable

lifeboat grabline detached

steel shell wasted

detached rudder

wasted stern frame of lifeboat

steel frame wasted

not water-tight window of lifeboat

canopy cracked due to accident

seized linkage

cradle holed

cradle bent

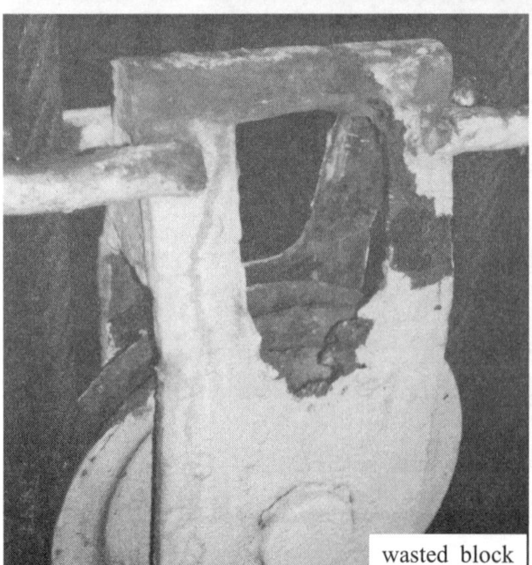

wasted block

wasted/holed foundation of davit

wasted winch foundation

- **LifeRaft**

[liferaft container tied with tapes]

(note: Some types are designed / approved to be tied with tapes. Such tapes are cut automatically on inflation)

- **Load Line**

freeboard mark not painted

air pipe head opened

air pipe head wasted

air pipe head wasted

Appendix B : Sample Checklist

SECTION A - SHIP'S CERTIFICATES

NO	NAME OF CERTIFICATE	DOCUMENT Ref.	NOTE
1	Certificate Of Registry		
2	Certificate Of Class		
3	Certificate Of Seaworthiness	National Legislation	
4	Safety Construction Certificate	SOLAS 92 I/12(a)(ii)	
5	Safety Equipment Certificate	SOLAS 92 I/12(a)(iii)	
6	Safety Certificate	SOLAS 92 I/12(a)(iii)	
7	Safety Radio Certificate	SOLAS 92 I/12(a)(iv)	
8	Solas Exemption Certificate	SOLAS 92 I/12(a)(vi)	
9	International Load Line Certificate	LOAD LINE 66 art. 16	
10	Load Line Exemption Certificate	LOAD LINE 66 art. 6	
11	International Tonnage Certificate	International Tonnage Convention, 1969 art. 7	
12	Minimum Safe Manning Document	S74 CV R 14 IMO Res. A 890	
13	IOPP Certificate	MARPOL 73/78 Annex I reg.5	
14	Sewage Poll. Prev. Certificate (Ispp)	MARPOL 73/78 Annex IV reg.4	
15	NLS/IPPC Certificate MARPOL 73/78	Annex II reg.12, 12A	
16	ICOF / COF Chem IBC Code Section 1.5 / BCH Code Section 1.6	ICOF / COF gas IGC Code section 1.5 / GC Code and GC ex Code section 1.6	
17	Safety Management Certificate (ISM)	SOLAS 92 IX/4.3	
18	Document Of Compliance (ISM)	SOLAS 92 IX/4.1	
19	Document Of Compliance For Dangerous Goods	SOLAS 92 II-2/54.3	
20	Certificate Of Compliance For Carriage Of Grain	SOLAS 92 VI/9 and IMO Res. MSC.23(59) (Grain Code) section 3	
21	Certificate Of Compliance For Carriage Of Solid Bulk Cargoes	BC Code IMO Code of Safe Practice for Solid Bulk Cargoes	
22	Cargo Gear Booklet	ILO Convention 152 art. 25	
23	ILO Crew Accommodation	ILO Conventions 92 and 133	
24	International Ship Security Certificate (Issc)	S02 CXI-2	
25	Continuous Synopsis Record (CSR)	S02 CXI-1 R 5	
26	Radio Station License	ITU RR	

SECTION B - EXPLANATION OF SHIP'S CERTIFICATES & MANUELS

GENERAL CARGO	
International Tonnage Certificate (1969) An International Tonnage Certificate (1969) shall be issued to every ship, the gross and net tonnage of which have been determined in accordance with the Convention.	Tonnage Convention, article 7
International Load Line Certificate An International Load Line Certificate shall be issued under the provisions of the International Convention on Load Lines, 1966, to every ship which has been surveyed and marked in accordance with the Convention or the Convention as modified by the 1988 LL Protocol, as appropriate.	LL Convention, article 16; 1988 LL Protocol, article 18
International Load Line Exemption Certificate An International Load Line Exemption Certificate shall be issued to any ship to which an exemption has been granted under and in accordance with article 6 of the Load Line Convention or the Convention as modified by the 1988 LL Protocol, as appropriate.	LL Convention, article 6; 1988 LL Protocol, article 18
Minimum safe manning document Every ship to which chapter I of the Convention applies shall be provided with an appropriate safe manning document or equivalent issued by the Administration as evidence of the minimum safe manning.	SOLAS 1974 (2000 amendments), regulation V/14.2
International Oil Pollution Prevention Certificate An international Oil Pollution Prevention Certificate shall be issued, after survey in accordance with regulation 4 of Annex I of MARPOL 73/78, to any oil tanker of 150 gross tonnage and above and any other ship of 400 gross tonnage and above which is engaged in voyages to ports or offshore terminals under the jurisdiction of other Parties to MARPOL 73/78. The certificate is supplemented with a Record of Construction and Equipment for Ships other than Oil Tankers (Form A) or a Record of Construction and Equipment for Oil Tankers (Form B), as appropriate.*	MARPOL 73/78, Annex I, regulation 5

International Sewage Pollution Prevention Certificate An International Sewage Pollution Prevention Certificate shall be issued, after an initial or renewal survey in accordance with the provisions of regulation 4 of Annex IV of MARPOL 73/78, to any ship which is required to comply with the provisions of that Annex and is engaged in voyages to ports or offshore terminals under the jurisdiction of other Parties to the Convention.	MARPOL 73/78, Annex IV, regulation 5; MEPC/Circ.408
Voyage data recorder system - certificate of compliance The voyage data recorder system, including all sensors, shall be subjected to an annual performance test. The test shall be conducted by an approved testing or servicing facility to verify the accuracy, duration and recoverability of the recorded data. In addition, tests and inspections shall be conducted to determine the serviceability of all protective enclosures and devices fitted to aid location. A copy of the certificate of compliance issued by the testing facility, stating the date of compliance and the applicable performance standards, shall be retained on board the ship.	SOLAS 1974, regulation V/18.8
Document of Compliance A document of compliance shall be issued to every company which complies with the requirements of the ISM Code. A copy of the document shall be kept on board.	SOLAS 1974, regulation IX/4; ISM Code, paragraph 13
Safety Management Certificate A Safety Management Certificate shall be issued to every ship by the Administration or an organization recognized by the Administration. The Administration or an organization recognized by it shall, before issuing the Safety Management Certificate, verify that the company and its shipboard management operate in accordance with the approved safety management system.	SOLAS 1974, regulation IX/4; ISM Code, paragraph 13
International Ship Security Certificate (ISSC) or Interim International Ship Security Certificate An International Ship Security Certificate (ISSC) shall be issued to every ship by the Administration or an organization recognized by it to verify that the ship complies with the maritime security provisions of SOLAS chapter XI-2 and part A of the ISPS Code. An interim ISSC may be issued under the ISPS Code part A, section 19.4.	SOLAS 1974 (2002 amendments), regulation XI-2/9.1.1; ISPS Code part A, section 19 and appendices.

Continuous Synopsis Record (CSR) Every ship to which chapter I of the Convention applies shall be issued with a Continuous Synopsis Record. The Continuous Synopsis Record provides an on-board record of the history of the ship with respect to the information recorded therein.	SOLAS 1974 (2002 amendments), regulation XI-1/5
Cargo Ship Safety Construction Certificate A certificate called a Cargo Ship Safety Construction Certificate shall be issued after survey to a cargo ship of 500 gross tonnage and over which satisfies the requirements for cargo ships on survey, set out in regulation I/10 of SOLAS 1974, and complies with the applicable requirements of chapters II-1 and II-2, other than those relating to fire-extinguishing appliances and fire control plans.	SOLAS 1974, regulation I/12, as amended by the GMDSS amendments; 1988 SOLAS Protocol, regulation I/12
Cargo Ship Safety Equipment Certificate A certificate called a Cargo Ship Safety Equipment Certificate shall be issued after survey to a cargo ship of 500 gross tonnage and over which complies with the relevant requirements of chapters II-1 and II-2 and III and any other relevant requirements of SOLAS 1974. A Record of Equipment for the Cargo Ship Safety Equipment Certificate (Form E) shall be permanently attached.	SOLAS 1974, regulation I/12, as amended by the GMDSS amendments; 1988 SOLAS Protocol, regulation I/12 (2000 amendments), appendix
Cargo Ship Safety Radio Certificate A certificate called a Cargo Ship Safety Radio Certificate shall be issued after survey to a cargo ship of 300 gross tonnage and over, fitted with a radio installation, including those used in life-saving appliances, which complies with the requirements of chapters III and IV and any other relevant requirements of SOLAS 1974. A Record of Equipment for the Cargo Ship Safety Radio Certificate (Form R) shall be permanently attached.	SOLAS 1974, regulation I/12, as amended by the GMDSS amendments; 1988 SOLAS Protocol, regulation I/12
Cargo Ship Safety Certificate A certificate called a Cargo Ship Safety Certificate may be issued after survey to a cargo ship which complies with the relevant requirements of chapters II-1, II-2, III, IV and V and other relevant requirements of SOLAS 1974 as modified by the 1988 SOLAS Protocol, as an alternative to the above cargo ship safety certificates. A Record of Equipment for the Cargo Ship Safety Certificate (Form C) shall be permanently attached.	1988 SOLAS Protocol, regulation I/12 (2000 amendments),

Exemption Certificate When an exemption is granted to a ship under and in accordance with the provisions of SOLAS 1974, a certificate called an Exemption Certificate shall be issued in addition to the certificates listed above.	SOLAS 1974, regulation I/12 ; 1988 SOLAS Protocol, regulation I/12
Intact stability booklet Every passenger ship regardless of size and every cargo ship of 24 metres and over shall be inclined on completion and the elements of their stability determined. The master shall be supplied with a Stability Booklet containing such information as is necessary to enable him, by rapid and simple procedures, to obtain accurate guidance as to the stability of the ship under varying conditions of loading. For bulk carriers, the information required in a bulk carrier booklet may be contained in the stability booklet.	SOLAS 1974, regulations II-1/22 and II-1/25-8; 1988 LL Protocol, regulation 10
Damage control plans and booklets On passenger and cargo ships, there shall be permanently exhibited plans showing clearly for each deck and hold the boundaries of the watertight compartments, the openings therein with the means of closure and position of any controls thereof, and the arrangements for the correction of any list due to flooding. Booklets containing the aforementioned information shall be made available to the officers of the ship.	SOLAS 1974, regulations II-1/23, 23-1, 25-8; MSC/Circ.919
Fire safety training manual A training manual shall be written in the working languago of tho ship and shall be provided in each crew mess room and recreation room or in each crew cabin. The manual shall contain the instructions and information required in regulation II-2/15.2.3.4. Part of such information may be provided in the form of audio-visual aids in lieu of the manual.	SOLAS 1974 (2000 amendments), regulation II-2/15.2.3
Fire Control plan/booklet General arrangement plans shall be permanently exhibited for the guidance of the ship's officers, showing clearly for each deck the control stations, the various fire sections together with particulars of the fire detection and fire alarm systems and the fire-extinguishing appliances etc. Alternatlvely, at the discretion of the Administration, the aforementioned details may be set out in	SOLAS 1974 (2000 amendments), regulation II-2/15.2.4

a booklet, a copy of which shall be supplied to each officer, and one copy shall at all times be available on board in an accessible position. Plans and booklets shall be kept up to date; any alterations shall be recorded as soon as practicable. A dupplicate set of fire control plans or a booklet containing such plans shall be permanently stored in a prominently marked weathertight enclosure outside the deckhouse for the assistance of shore-side fire-fighting personnel.	
On board training and drills record Fire drills shall be conducted and recorded in accordance with the provisions of regulations III/19.3 and III/19.5.	SOLAS 1974 (2000 amendments), regulation II-2/15.2.2.5
Fire safety operational booklet The fire safety operational booklet shall contain the necessary information and instructions for the safe operation of the ship and cargo handling operations in relation to fire safety. The booklet shall be written in the working language of the ship and be provided in each crew mess room and recreation room or in each crew cabin. The booklet may be combined with the fire safety training manuals required in regulation II-2/15.2.3.	SOLAS 1974 (2000 amendments), regulation II-2/16.2
Certificates for masters, officers or ratings Certificates for masters, officers or ratings shall be issued to those candidates who, to the satisfaction of the Administration, meet the requirements for service, age, medical fitness, training, qualifications and examinations in accordance with the provisions of the STCW Code annexed to the International Convention on Standards of Training, Certification and Watchkeeping for Seafarers, 1978. Formats of certificates are given in section A-I/2 of the STCW Code. Certificates must be kept available in their original form on board the ships on which the holder is serving.	STCW 1978, article VI, regulation I/2; STCW Code, section A-I/2
Oil Record Book Every oil tanker of 150 gross tonnage and above and every ship of 400 gross tonnage and above other than an oil tanker shall be provided with an Oil Record Book, Part I (Machinery space operations). Everyoil tanker of 150 gross tonnage and above shall also be provided with an Oil Record Book, Part II (Cargo/ballast operations).	MARPOL 73/78, Annex I, regulation 20

Shipboard Oil Pollution Emergency Plan Every oil tanker of 150 gross tonnage and above and every ship other than an oil tanker of 400 gross tonnage and above shall carry on board a Shipboard Oil Pollution Emergency Plan approved by the Administration.	MARPOL 73/78, Annex I, regulation 26
Garbage Management Plan Every ship of 400 gross tonnage and above and every ship which is certified to carry 15 persons or more shall carry a garbage management plan which the crew shall follow.	MARPOL 73/78, Annex V, regulation 9
Garbage Record Book Every ship of 400 gross tonnage and above and every ship which is certified to carry 15 persons or more engaged in voyages to ports or offshore terminals under the jurisdiction of other Parties to the Convention and every fixed and floating platform engaged in exploration and exploitation of the sea-bed shall be provided with a Garbage Record Book.	MARPOL 73/78, Annex V, regulation 9
Cargo Securing Manual All cargoes, other than solid and liquid bulk cargoes, cargo units and cargo transport units, shall be loaded, stowed and secured throughout the voyage in accordance with the Cargo Securing Manual approved by the Administration. In ships with ro-ro spaces, as defined in regulation II-2/3.41, all securing of such cargoes, cargo units and cargo transport units, in accordance with the Cargo Securing Manual, shall be completed before the ship leaves the berth. The Cargo Securing Manual is required on all types of ships engaged in the carriage of all cargoes other than solid and liquid bulk cargoes, which shall be drawn up to a standard at least equivalent to the guidelines developed by the Organization.	SOLAS 1974 (2002 amendments), regulations VI/5. .6 6 and VII/5; MSC/Circ.745
Ship Security Plan and associated records Each ship shall carry on board a ship security plan approved by the Administration. The plan shall make provisions for the three security levels as defined in part A of the ISPS Code. Records of the following activities addressed in the ship security plan shall be kept on board for at least theminimum period specified by the Administration:	SOLAS 1974 (2002 amendments), regulation XI-2/9; ISPS Code part A, sections 9 and 10

.1 training, drills and exercises; .2 security threats and security incidents; .3 breaches of security; .4 changes in security level; .5 communications relating to the direct security of the ship such as specific threats to the ship or to port facilities the ship is, or has been, in; .6 internal audits and reviews of security activities; .7 periodic review of the ship security assessment; .8 periodic review of the ship security plan; .9 implementation of any amendments to the plan; and .10 maintenance, calibration and testing of any security equipment provided on board, including testing of the ship security alert system.	
Cargo Information The shipper shall provide the master or his representative with appropriate information, confirmed in writing, on the cargo, in advance of loading. In bulk carriers, the density of the cargo shall be provided in the above information.	SOLAS 1974, regulations VI/2 and XII/10; MSC/Circ.663

SECTION C - MANNING

No	Item	Ref.	Guidelines
1	Ship manned in accordance with the Safe Manning Document	**All ships** V/13	**All ships: ≥ 500 GT** shall be provided with an appropriate safe manning document or equivalent issued by the Administration as evidence of the minimum safe manning considered necessary to ensure safety of life at sea.
2	Certificates of the officers	**STCW** II/2, III/2	**All ships: ≥ 200 GT** shall be crewed with master, chief mate, chief engineer and second engineer holding an appropriate certificate for their rank. The certificates shall be issued under the provisions of the STCW Convention 1978 by the Flag Administration and shall specify any limitation imposed.
3	Certificates of the officers in charge of a navigational / engine watch	**STCW** II/4, III/4	**All ships: ≥ 200 GT** shall have navigational / engine watches performed by officers holding an appropriate certificate for their rank.The certificates shall be issued under the provisions of the STCW Convention 1978 by the Flag Administration and shall specify any limitation imposed.
4	Certificated lifeboatmen	**All ships** III/10	**All ships:** deck officers or certificated persons as per Safe Manning Document for operating the survival craft and launching arrangement. One shall be in charge for each survival craft and have a list of its crew. A person capable of operating the engine to be assigned to each motorised survival craft.

SECTION D - SAFETY DOCUMENTATION AND CREW FAMILIARISATION

No	Item	Ref.	Guidelines
1	Stability booklet, associated stability plans and information	all ships II-1/22 S 92 II-1/25-8 ILLC/10 M I/25(5) IBC/2 IGC/2	**All ships:** to be supplied with stability information approved by the Flag Administration.
2	Grain loading manual	all ships VI/9	**All ships:** carrying grain in bulk shall be provided with information regarding ship's stability and grain loading in printed booklet, approved by the Flag Administration (see also IMO Grain Code section 6)
3	Loading manual / electronic loading instrument operation manual	ILLC/10 S92 XII/11	**All ships:** to be supplied with approved information for the loading and ballasting of the ship avoiding creation of unacceptable stresses **All ships:** have to be provided with an approved loading manual, except those of Category II less than 90 m in length in which the deadweight does not exceed 30% of the displacement at the summer loadline draught **All ships of Category I:** have to be provided with an approved loading instruments **All bulk carriers, ore carriers and combination carriers:** having L $\geq$150 m are to be provided with an approved loading manual and an approved computer-based loading instrument

No	Item	Ref.	Guidelines
4	Damage control plans	**S 92** II-1/23-1	**S 92 (01.02.92) (applicable only to dry cargo ships):** plans permanently exhibited or available on the bridge showing boundaries and openings of watertight compartments, means of closure and position of controls, arrangements for correction of list due to flooding.
5	Manoeuvring information	**S 81** II-1/28	**S 81:** stopping times, ship headings and distances recorded on trials, performances with inoperative propellers (multiple propeller ships) available to the master.
6	Instructions for maintenance and operation of all installations/equipment for fighting and containment of fire: readily available, language understood by crew	**all ships** II-2/20	**All ships:** to be kept under one cover, readily available in an accessible position.
7	Training manual for lifesaving appliances: provided in each crew messroom and recreation room or in each crew cabin	**all ships** III/18.2, 51	**All ships:** to be provided with the manual / video concerning: donning of lifejackets and immersion suits, muster, boarding and launching the survival craft, illumination in launching areas, use of survival and detection equipment, use of radio equipment, emergency repairs of lifesaving appliances.
8	Instructions for on-board maintenance of lifesaving appliances	**all ships** III/19.3, 52	**All ships:** to include: checklist for lifesaving appliances inspection, maintenance and repair instructions, schedule of maintenance, lubricating points, replaceable parts, spare parts, log for records of inspections and maintenance.

No	Item	Ref.	Guidelines
9	Operating instructions for steering gear posted: navigating bridge and steering gear compartment	**all ships** V/19-2	**All ships:** block diagram showing the change-over procedures for remote steering gear control systems and steering gear power units.
10	Oil Record Book	**M I**/20	**All ships:** to be provided with an Oil Record Book Part I (Machinery space operations) and all oil tankers to be provided also with Part II (Cargo/ballast operations). **Entries in Part I:** ballasting or cleaning oil fuel tanks, discharge of dirty ballast / cleaning water from oil fuel tanks, disposal of sludge, discharge overboard / disposal of bilge water accumulated in machinery spaces. **Entries in Part II:** loading/unloading of oil cargo, internal transfer of oil cargo during voyage, ballasting of cargo tanks / CBT, cleaning of cargo tanks including COW, discharge of ballast (except SBT), discharge from slop tanks, closing of slop tanks valves and CBT/cargo tanks separation valves after slop tanks discharge operations, disposal of residues. Entries to be at least in English or French, signed by the responsible officer. Oil Record Book to be available at all times and kept for three years after the last entry.
11	S.O.P.E.P	**M I**/26	**All ships:** shall have this plan approved by the Flag Administration.

No	Item	Ref.	Guidelines
12	Charts and publications for intended voyage (charts, coast pilot, sailing directions, light list, tide tables, tidal current tables): available and updated	**all ships** V/20	**All ships:** shall carry adequate and up-to-date charts, sailing directions, lists of lights, notices to mariners, tide tables and all other nautical publications necessary for the intended voyage.
13	Fire plans: permanently exhibited, permanently stored in weathertight containers outside deckhouse, updated, language understood by crew	**all ships** II-2/20	**All ships:** permanently exhibited, showing clearly, for each deck: control stations, A and B divisions, fire detection and alarm system, sprinkler, fire-extinguishing appliances, means of access to compartments, ventilating system including fans control positions, dampers, relationship between fans and ventilated spaces. Alternatively, booklets shall be supplied to each officer and one copy shall be available in an accessible position. They shall be written using the Flag State language and English or French if different. A duplicate set of fire plans or booklets shall be permanently stored in a prominently marked weathertight enclosure outside the deckhouse.
14	Posters or signs provided on or in the vicinity of survival craft and their launching control	**all ships** III/9	**All ships:** posters or signs shall be provided on or in the vicinity of survival craft and their launching control and shall: −illustrate the purpose of controls and the procedures for operating the appliance and give relevant instructions or warnings −be easily seen under emergency lighting conditions

No	Item	Ref.	Guidelines
			–use symbols in accordance with IMO recommendations (refer to IMO Res. A.603(15))
15	International code of signals	**all ships** V/21	**All ships:** required to carry radio installations, shall have a copy of the international code of signals.
16	Muster list: crew members duties shown, posted in conspicuous places, language understood by crew	**all ships** III/8, 53	**All ships:** to be exhibited in conspicuous places including bridge, engine room and crew accommodation spaces. It shall specify details on the emergency alarm signal and consequent actions to be taken. It shall include crew duties: operation of watertight and fire closing appliances; equipping, preparation and launching of survival craft; use of communication equipment; manning of fire parties; special duties for fire fighting; officers responsible for keeping lifesaving appliances ready and in good condition; substitutes of key persons.
17	Cargo securing manual S92 VI/5.6	**S92** VI/5.6 **S92** VII/5.6	**All ships:** cargo units, including containers, shall be loaded, stowed and secured in accordance with the Cargo Securing Manual
18	Procedures and arrangements manual	**M II**/5	**All chemical tankers:** shall have this manual approved by the Flag Administration.
19	Cargo Record Book	**M II**/9	**All chemical tankers:** to be provided with a Cargo Record Book as part of the official ship's log-book or in a form according to MARPOL standard. Entries in the book: loading/ unloading of cargo, internal transfer of

No	Item	Ref.	Guidelines
			cargo during voyage, ballasting of cargo tanks, cleaning of cargo tanks, discharge of ballast from cargo tanks, disposal of residues to reception facilities, discharge into the sea or removal by ventilation of residues in accordance with M II/5. Entries in the book shall be at least in English or French, signed by the responsible officer. Oil Record Book to be available at all times and kept for three years after the last entry.
20	Information for the safe carriage of the products	**IBC**/16.2.3 **BCH**/5.2.2 **IGC**/18	**All chemical tankers and gas carriers:** to be provided with information for the products being carried. including a stowage plan, physical and chemical properties, actions against spills, actions against personal contact, fire-fighting, cargo transfer, cleaning, gasfreeing, ballasting.
21	Enhanced survey programme report file	**all ships** XI/2	**All ships (bulk carriers and oil tankers):** subject to an enhanced survey programme shall have on board, for their lifetime, a survey report file consisting of: reports of structural surveys, condition evaluation reports and thickness measurements reports.
22	Instructions for stowage of bulk cargoes	**All ships** VI/7.2	**All cargo ships carrying bulk cargoes:** to be provided with a booklet including following information: stability data, ballasting and deballasting rates and capacities, maximum allowable load per unit surface area of tank top plating,

No	Item	Ref.	Guidelines
			maximum allowable load per hold, general loading and unloading instructions, any special restriction imposed, maximum permissible forces and moments during loading, unloading and voyage.
23	Loading and unloading sequences	**S 92** VI/7.2	**All ships (single skin bulk carriers of 150 m length and above):** contracted for construction before 01/07/1998: to be provided with an approved manual with typical loading sequences
24	Inert gas system instruction manual	**all ships** II-2/62.21	**All ships:** to be provided with manual concerning operations, safety, maintenance requirements, occupational health hazards and procedures for faults or failures connected to the use of the IGS.
25	Dedicated CBT operation manual	**M I**/13A(4)	**All oil tankers:** provided with a dedicated CBT system, shall have this manual approved by the Flag Administration.
26	COW operations and equipment manual	**M I**/13B(5)	**All oil tankers:** provided with a COW system, shall have this manual approved by the Flag Administration.
27	Operational procedures for special ballast arrangements	**M I**/13D(1)(a)	**All existing oil tankers:** having special ballast arrangements, shall have these procedures approved by the Flag Administration.
28	Operations manual for the oil discharge monitoring and control system	**M I**/15(3)(c)	**All oil tankers:** provided with an ODMCS, shall have this manual approved by the Flag Administration.

No	Item	Ref.	Guidelines
29	Witness an abandon ship drill. Verify that no language problems exist and the crew is properly mustered and familiar with their duties.	**all ships** III/18.3	**All ships:** each member of the crew shall participate in at least one abandon ship drill every month. This drill shall take place within 24 h of the ship leaving a port if more than 25% of the crew have not participated in abandon ship on board that particular ship in the previous month. An abandon ship drill shall include: –summoning of crew to muster stations with emergency alarm –verifying that lifejackets are correctly donned –lowering of at least one lifeboat (each lifeboat to be launched and manoeuvred in the water every 3 months). –starting and operating the lifeboat engine –operation of davits used for launching liferafts.
30	Witness a fire drill. Verify that no language problems exist, the crew is familiar with the information given in the fire control plan/booklet and with their duties, fire teams are properly dressed and equipped, communication and coordination exist between bridge and fire team.	**all ships** III/18.3	

SECTION E - LOGBOOK ENTRIES

No	Item	Ref.	Guidelines
1	Steering gear test/drills (12 hours before departure)	**all ships** V/19-2	**All ships:** within 12 hours before departure steering gear to be checked and tested. Items concerned: main and auxiliary steering gears, remote control system, bridge steering position, emergency power supply, rudder angle indicators, remote control power failure alarms, steering gear power failure alarms, automatic equipment. Checks required: full movement of the rudder, visual inspection of the steering gear, operation of means of communication.
2	Three-monthly emergency steering drills	**all ships** V/19-2	**All ships:** every three months. Practice of emergency steering procedures: direct control from steering gear compartment, communication, alternative power supplies.
3	Monthly abandon ship drills	**all ships** III/18.3, 5	**All ships:** one drill each month. Each lifeboat and rescue boat launched every three months.
4	Monthly fire drills	**all ships** III/18.3, 5	**All ships:** one drill each month.
5	On-board training in use of lifesaving equipment	**all ships** III/18.4, 5	**All ships:** each crew member to be trained in the use of lifesaving appliances not later than 2 weeks after he joins the ship. Training in use of liferaft davits each 4 months.
6	Weekly visual inspection of survival craft / rescue boat and launching appliances	**all ships** III/19.6.1	**All ships:** weekly inspections compulsory, logbook entry recommended

No	Item	Ref.	Guidelines
7	Weekly operation of lifeboat / rescue boat engines	**all ships** III/19.6.2	**All ships:** weekly inspections compulsory, logbook entry recommended
8	Weekly testing of the general emergency alarm	**all ships** III/19.6.3	**All ships:** weekly inspections compulsory, logbook entry recommended
9	Monthly inspection of lifesaving appliances and lifeboat equipment using the relevant checklist	**all ships** III/19.7	**All ships:** every month, using the checklist provided in the instructions for on-board maintenance.
10	Monthly checking of breathing apparatus	**all chem.** **IBC**/14.2.6 **BCH**/3.16.8	**All chemical tankers:** a responsible officer should inspect each breathing apparatus once a month.

SECTION F - CREW SAFETY AND HEALTH

No	Item	Ref.	Guidelines
1	Rails, guards, protective clothing and equipment, warning signs posted in crew work areas	**ILO** **S81** II-1/36 **S92** II-1/36	**S81, S92:** ear protectors provided for personnel working in excessively noisy spaces.

SECTION G - SAFETY OF NAVIGATION

No	Item	Ref.	Guidelines
1	Daylight signalling lamp and emergency batteries	**all ships** V/11	**All ships: ≥ 150 GT** to be provided with an efficient daylight signalling lamp connected also to the emergency system.
2	Magnetic compasses: liquid level satisfactory, deviation table provided, speaking tube between magnetic compass and bridge	**all ships** V/12(b)(i), (iii)	**All ships: ≥ 150** GT to be provided with a standard magnetic compass + a steering magnetic compass unless information from standard m.c. are readable by the helmsman at the steering position. Adequate means of communication between standard and navigation control positions to be provided. A spare magnetic compass interchangeable with the standard m.c. to be provided unless a steering m.c. or a gyrocompass is provided. All ships: each a.m. magnetic compass to be adjusted and its table of residual deviations to be available at all times.
3	Gyro compass	**all ships** V/12(d), (e)	**All ships:** built on or after 1.9.84, and ≥ 500 GT to be fitted with a master gyrocompass readable by the helmsman; built on or after 1.9.84, and ≥ 1600 GT to be fitted with gyrocompass repeater(s) for taking bearings over an arc of 360°; built before 1.9.84, and ≥ 1600 GT to be fitted with a master gyrocompass readable by the helmsman and with gyrocompass repeater(s) for taking bearings over an arc of 360°.
4	Emergency steering information	**all ships** V/12(f)	**All ships:** provided with a telephone or other means of communication for relaying heading information. Ships built

No	Item	Ref.	Guidelines
			on or after 1.2.92 and ≥ 500 GT: to be provided for arrangements for supplying visual compass readings to the emergency steering position.
5	Radar(s)	**all ships** V/12(g), (h)	**All ships:** ≥ 300 GT shall be fitted with a radar installation capable of operating in the 9 Ghz frequency band. Ships ≥ 10000 GT shall be fitted with two radar installations, at least one being capable of operating in the 9 Ghz frequency band.
6	Radar(s) plotting facilities	**all ships** V/12(i)	**All ships:** required to have a radar installation shall have facilities for plotting radar readings (at least as effective as a reflection plotter for ships built on or after 1.9.84 and ≥ 1600 GT).
7	Automatic Radar Plotting Aid (ARPA)	**all ships** V/12(j)	**All ships: ≥ 10000 GT** built on or after 1.9.84 shall be fitted with an ARPA Tankers: ≥10000 GT built before 1.9.84 shall be fitted with an ARPA All ships (not tankers): ≥15000 GT built before 1.9.84 shall be fitted with an ARPA
8	Echo-sounding device: paper and ink provided	**all ships** V/12(k)	**All ships: ≥ 500 GT** (built on or after 25.5.80) and ≥ 1600 GT (built before 25.5.80) shall be fitted with an echo-sounding device.
9	Speed and distance indicator	**all ships** V/12(l)	**All ships:** built on or after 1.9.84 and ≥ 500 GT shall be fitted with a device to indicate speed and distance. Ships required to be fitted with an ARPA shall be fitted with a device to indicate speed and distance through the water.
10	Rudder angle indicator	**all ships** V/12(m)	**All ships: ≥ 500 GT** (built on or after 1.9.84) and ≥ 1600 GT (built before 1.9.84) shall be fitted with an indicator

No	Item	Ref.	Guidelines
			showing the rudder angle readable from the conning position.
11	Propellers rate of revolution indicator	**all ships** V/12(m)	**All ships: ≥ 500 GT** (built on or after 1.9.84) and ≥ 1600 GT (built before 1.9.84) shall be fitted with indicators showing the rate of revolution of each propeller readable from the conning position.
12	Variable pitch propeller/lateral thrust propellers pitch and operational mode indicator	**all ships** V/12(m)	**All ships: ≥ 500 GT** (built on or after 1.9.84) and ≥ 1600 GT (built before 1.9.84) shall be fitted with indicators showing the pitch and operational mode of variable pitch propellers and / or lateral thrust propellers readable from the conning position.
13	Rate of turn indicator	**all ships** V/12(n)	**All ships:** built on or after 1.9.84 and ≥ 100000 GT shall be fitted with a rate of turn indicator.
14	Radio homing device	**all ships** V/12(q)	**All ships: ≥ 1600 GT** built on or after 25.5.80 but before 1.2.95 shall be fitted with a radio equipment for homing on the radiotelephone distress frequency (2182 kHz) until 1.2.99.
15	Records of periodical tests performed on board for items not tested	**all ships** V/12	**All ships:** when an equipment cannot be tested due to the condition in which the survey is carried out (e.g. speed indicator, ship alongside), the relevant records of periodical inspections must be checked.
16	Pilot ladders and hoists/pilot transfer arrangements, lighting, heaving line and lifebuoy with self-igniting light	**all ships** V/17	**All ships:** efficient, kept clean and in good order. Gateway, handholds, stanchions for safe passage on to /into / off the ship. Proper lighting, lifebuoy equipped with self-igniting light and heaving light ready for use. Mechanical

No	Item	Ref.	Guidelines
			pilot hoist (where provided) of type approved and efficient.
17	Lights and shapes	**COL** 20-24, 27-30	**All ships:** (when power driven) shall show a fwd masthead white light, an aft masthead white light (L ≥ 50 m), sidelights (green stbd, red port) and a white sternlight. When under towing shall show a towing yellow light, two masthead white lights in a vertical line (three and a diamond shape when the length of the tow exceeds 200 m), sidelights (green stbd, red port) and a white sternlight. When not under command shall show two all-round red lights, two balls, sidelights (green stbd, red port) and a white sternlight. When anchored shall show a fwd all-round white light or a ball and an aft allround white light at a lower level. Vertical positioning of lights: fwd masthead light at least 6 m (or ship's breadth if greater) above the hull (need not to exceed 12 m). When two mastheads are carried, aft one higher at least 4,5 m than fwd one. Aft one above the fwd one when seen 1000 m from the stem at sea-level in all normal conditions of trim. Sidelights not above 3/4 height above hull of fwd masthead light. Horizontal positioning: of lights: two masthead lights spaced more than 0,5 L (need not to exceed 100 m). Fwd masthead light not more than 0,25 L from the stem. Sidelights not forward the fwd masthead light.
18	ship whistle, bell and gong	**COL** 33	**All ships:** shall be provided with a whistle and a bell. If L≥100m shall be in addition provided with a gong

SECTION H - HULL: STRUCTURAL INTEGRITY AND ARRANGEMENTS

H-1 - STRUCTURAL INTEGRITY

No	Item	Ref.	Guidelines
1	**General examination of the hull:** fractures, wastage, pitting or damage to the extent that may impair ship's seaworthiness, doublers fitted without being recorded	**ILLC**/1	General strength of the hull to be sufficient for the draught corresponding to the freeboard assigned.
2	Freeboard marks properly marked and painted. Check positions of deck line and load line.	**ILLC**/8 **IACS** LL4 int.	The ring, lines and letters shall be painted in white or yellow on a dark ground or in black on a light ground. They shall also permanently marked on the sides of the ship. Permanently marked is considered to include welding of the marks.
3	Weathertight doors: permanently and strongly attached to the bulkhead gaskets and clamping devices provided and efficient operated from both sides of the bulkhead.	**ILLC**/12	All access openings in bulkheads at ends of enclosed superstructures shall be fitted with doors of steel or other equivalent material and framed, stiffened and fitted so that the whole structure is of equivalent strength to the unpierced bulkhead and weathertight when closed.
4	Cargo and other hatchways and coamings	**ILLC**/14,15,16	General strength of coamings and covers satisfactory. Condition of means for securing and maintaining weathertightness satisfactory. Test for tightness may be required at surveyor's satisfaction.
5	Machinery space openings: weathertightness, covers, casings and coamings	**ILLC**/17	Machinery space openings shall be properly framed and efficiently enclosed by steel casings. Access openings in such casings shall be fitted with doors

No	Item	Ref.	Guidelines
			complying with the requirements of Reg. ILLC/12(1). Sills: h〉600mm (pos.1), h〉380mm (pos.2).
6	Manholes and flush scuttles: watertightness, covers and bolts	**ILLC**/18(1)	Manholes and flush scuttles shall be closed by substantial covers capable of being made watertight. Unless secured by closely spaced bolts, the covers shall be permanently attached.
7	Other openings in freeboard decks and associated deckhouse or companionway: weathertightness, bulkhead plating, doorways, gaskets and clamping devices	**ILLC**/18(2)	Openings in freeboard decks other than hatchways, machinery space openings manholes and flush scuttles shall be protected by an enclosed superstructure or by a deckhouse or companionway. Doorways in such deckhouses or companionways shall be fitted with doors complying with Reg. ILLC/12(1). Sills: h〉600mm (pos.1), h〉380mm (pos.2).
8	Ventilators: coamings, closing covers, gaskets, clamping devices	**ILLC**/19	Satisfactory strength of coamings. Efficient weathertight closing appliances provided and, in ships with length 〈 100 m, permanently attached (closing appliances not required where coamings extend more than 4,5 m above the deck (position 1) or 2,3 m (position 2). Coamings: h〉900mm (pos.1), h〉760mm (pos.2).
9	Air pipes: heads, means for closing the openings	**ILLC**/20	Satisfactory strength of pipes above deck. Efficient and permanently attached closing appliances provided. Height from the deck: h〉760mm (freeboard deck), h〉450mm (superstructure deck).
10	Watertight and structural integrity of openings below the freeboard deck	**ILLC**/21	Cargo ports and other similar openings in the sides of ships below the freeboard deck shall be fitted with doors

No	Item	Ref.	Guidelines
			so designed as to ensure watertightness and structural integrity commensurate with the surrounding shell plating.
11	Scuppers, inlets and discharges: non-return valves, positive means of closing (when fitted)	**ILLC**/22	Efficient and accessible means for preventing water from passing inboard provided for discharges from spaces below the freeboard deck and from superstructures fitted with weathertight doors. Where means for operating the positive action valve are provided, they shall be readily accessible and provided with an indicator showing whether the valve is open or closed. Scuppers and discharge pipes penetrating the shell either more than 450mm below the freeboard deck or less than 600mm above the summer load waterline shall be provided with a non-return valve at the shell. This valve may be omitted if the piping is of substantial thickness.
12	Sidescuttles	**ILLC**/23	Side scuttles to spaces below the freeboard deck or to spaces within enclosed superstructures shall be fitted with efficient hinged inside deadlights arranged so that they can be effectively closed and secured watertight.
13	Compliance of vertically corrugated bulkhead between cargo holds no1 and no 2	**S 92** XII/4-6-9	**All single side skin bulk carriers:** built before 01/07/98 and having L≥150 m carrying solid bulk cargoes having bulk density of 1,78 t/m^3 or above fitted with vertically corrugated bulkheads between cargo holds no 1 and 2, have to comply with minimum scantlings requirements.Compliance to be re-assessed at each Special Survey

No	Item	Ref.	Guidelines
14	Efficiency of means of detection of water ingress into cargo holds	S 92 XII/10	**All single side skin bulk carriers:** built before 01/07/98 and having L≥150 m carrying solid bulk cargoes having bulk density of 1,78 t/m^3 constructed with an insufficient number of transverse watertight bulkheads to withstand flooding of foremost cargo hold, have to be provided with high water level alarms in all cargo holds, or in cargo conveyor tunnels, giving an audible and visual alarm on the navigation bridge, and with detailed information on flooding scenarios and evacuation preparedness

H-2 - HULL ARRANGEMENTS

No	Item	Ref.	Guidelines
1	Guard-rails, bulwarks, walkways	**ILLC**/25	Efficient guard-rails or bulwarks are fitted on all exposed parts of the freeboard and superstructure decks. Satisfactory means (i.e.: guardrails, lifelines, gangways or underdeck passages) are provided for the protection of the crew in getting to and from their quarters, the machinery space and all the other parts used in the necessary work of the ship.
2	Arrangements for safe access to bow	**S92** II-1/3-3	**All oil tanker, chemical tanker, gas carrier:** not later than 01/07/01 to be provided with suitable means to gain safe access to the bow even in severe weather conditions, which can be an under-deck passageway , a permanent gangway fitted at or above the level of the superstructure deck or a permanent walkway fitted at freeboard deck level

No	Item	Ref.	Guidelines
3	Fitting for timber deck cargoes	**ILLC**/43, 44	**All ships:** fitted for carriage of timber deck cargoes shall be provided with bulwarks with freeing ports or specially strong construction rails. Ventilators shall be efficiently protected. Uprights to be provided when required by the nature of the timber. Spacing not to exceed 3 m. Strong angles or metal socks or similar to be provided for securing the uprights. Over-all lashings to be provided (spacing not to exceed 3 m). Eye-plates for lashings to be fitted to the sheer strake or deck stringer plate (3 m spacing). First eye-plate from a superstructure bulkhead within 2 m. Minimum lashing diameter 19 mm (if close-link chain) or equivalent if wire rope, fitted with sliphooks and turnbuckles, accessible at all times. Wire rope lashings to have a short length of long link chain to permit regulation. Additionally, guard-rails or lifelines for protection of crew shall be provided each side of the deck cargo. Vertical separation of lines/rails $\leq$ 330 mm, total height $\geq$ 1 m.
4	Local and remote control of watertight doors and relevant indicators	**S 92** II-1/23-1,25-9	**S 92 (01.02.92):** dry cargo ships: sliding and hinged doors in watertight bulkheads and shell doors which could lead to major flooding to be provided with indicators on the bridge. Local and remote (from bridge) control to be provided. Control to be operable in case of main power failure. Doors normally closed at sea to be provided only with indicator system and permanently marked "to be kept close at sea".

No	Item	Ref.	Guidelines
5	Electronic loading instrument	**ILLC**/10	The instrument is to be tested for accuracy, using the load cases in the operation manual approved by Class.
6	Emergency towing arrangements	**all ships** II-1/3-4	**All oil tankers, chemical tankers and gas carriers ≥ 20,000 dwt:** an emergency towing arrangement approved by the Administration is to be fitted at both ends of. Such equipment is to be fitted at the first scheduled dry-docking survey and in any case within 1.1.99.
7	Navigation bridge visibility	**all ships** V/22	**All ships: having L ≥ 45 m** to comply with minimum requirements relevant to the view of the sea surface from the conning position and to blind sectors. For existing ships, these requirements will be met where practicable; structural alterations need not to be required

SECTION I - MACHINERY AND ELECTRICAL ARRANGEMENTS

Guidelines: Underlined items compulsory also for existing tankers, chemical tankers and gas carriers ≥ 10000 GT.

No	Item	Ref.	Guidelines
1	Condition of non-metallic expansion joints below deepest load waterline	**All ships** II-1/26.9	**All ships:** non-metallic expansion joints in piping systems which penetrate the ship's side and both the penetration and the expansion joints are located below the deepest waterline to be inspected and replaced at intervals recommended by manufacturer
2	Bilge pump and bilge lines	**S 81** II-1/21 **S 92** II-1/21	**S 81, S 92:** draining provided for any watertight compartment. Two pumps to be provided (one may be driven by

No	Item	Ref.	Guidelines
			main engine), sanitary, ballast and general service pumps accepted as bilge pumps. Distribution boxes and manually operated to be accessible at any time. Passage of water from the sea or from a compartment to another to be prevented.
3	Machinery, boilers, pressure vessels arranged in way to minimise danger to crew for hot surfaces, moving parts, etc.	**S 81** II-1/26 **S 92** II-1/26	**S 81, S 92:** machinery protected as to reduce to a minimum any danger to persons on board, due regard being paid to moving parts, hot surfaces and other hazards.
4	Absence of fire risks in machinery spaces (no leakage, bilge/floors free of oil)	**S 81** II-1/15 **S 92** II-1/15	**S 81, S 92:** free standing oil fuel tanks in machinery spaces to be provided with an oil-tight spill tray with drain pipe to a spill oil tank. Precaution to be taken preventing oil escape from pumps, filters or heaters towards heated surfaces. Non-metallic pipes accepted only if approved type. UMS shall have screened pipes to avoid oil spray to heated surfaces. **S 81:** sounding pipes not to terminate where risks of ignition may occur. Flat glass level gauges with self-closing devices accepted. **S 92 (01.02.92):** sounding pipes not to terminate in machinery spaces unless an oil-level gauge is provided, distant from sources of ignition or screened, provided with self-closing device and self-closing small diameter check-cock. Flat glass level gauges with self-closing devices accepted.

No	Item	Ref.	Guidelines
5	Boilers	**S 81** II-1/32 **S 92** II-1/32	**S 81, S 92:** not less than two safety valves and two means of indicating water level (one to be a direct reading gauge). Automatic boilers: alarm and shut off for: low water level air supply failure, flame failure. Propulsion boilers: high water level alarm.
6	Main and emergency steering gear system	**S 60** II/29, 30 **S 74** II-1/29, 30 **S 81** II-1/29, 30 **S 92** II-1/29, 30	**S 60, S 74:** auxiliary system capable of steering at navigable speed and of being brought speedily into action. To be power operated where rudder stock diameter 〉 355,6 mm. Not required where main system is duplicated. Ships ≥ 5000 GT: powered also by emergency source, two circuits widely separated. Ships 〈 5000 GT: auxiliary powered by motor intended for other services accepted. **S 81, S 92:** capable of steering up to 35° each side. Auxiliary system to be provided (duplication of main accepted provided that a single failure on one unit does not impair the other). Control of main and auxiliary from bridge (independent) and steering room. Where rudder stock diameter 〉 230 mm, emergency supply is required. Steering to be powered also by emergency source, two circuits widely separated. Ships 〈 1600 GT: auxiliary powered by motor intended for other services accepted. Tankers ≥ 10000 GT and other ships ≥ 70000 GT: duplication compulsory.

No	Item	Ref.	Guidelines
7	Steering gear room provided with handrails and gratings or other non-slip surfaces	**S 81** II-1/29	**S 81, S 92:** readily accessible, separated from machinery spaces, provided with railings and non-slip surfaces.
8	Correspondence of bridge and local rudder angle indicators	**S 60** II/29, 30 **S 74** II-1/29, 30 **S 81** II-1/29, 30 **S 92** II-1/29, 30	**S 60, S 74:** position indicated at the principal steering station. **S 81, S 92:** communication system to be tested. Indication provided both in bridge and steering room.
9	Steering gear alarms and indicators: –power failure (*) –low level for hydraulic fluid reservoir (*) –electric overload (*) also existing tankers/gas carriers	**S 60** II/30 **S 74** II/30 **S 81** II-1/29, 30 **S 92** II-1/29, 30	**S 60, S 74:** running indication of the electric motors in a suitable location. **S 81, S 92:** power failure to control system (bridge), low level of hydraulic fluid (bridge and E.R.). Running indication of the electric motors (bridge and suitable machinery control position). Electric overload in engine control room.
10	Operating instructions (block diagram) are permanently displayed on the navigation bridge and steering gear compartment	**all ships** V/19.2	**all ships:** simple operating instructions with a block diagram showing the change-over procedures for remote steering gear control system and steering gear power units shall be permanently displayed on the navigating bridge and in the steering gear compartment.
11	Ventilation for machinery spaces	**S 60** II/54, 69 **S 74** II-2/54 **S 81** II-2/48 **S 92** II-2/48	**S 60, S 74:** engine room ventilation capable of being stopped from the outside. Annular spaces around funnels closing appliances. **S 81, S 92:** additionally, ventilation of Cat. A machinery spaces not to pass through accommodation, service spaces, control stations unless provided with fire damper or A-60 insulated through the said spaces.

No	Item	Ref.	Guidelines
12	Engine room telegraph, secondary means of communication, communication with any other control position	**S 60** II/33 **S 74** II-1/32 **S 81** II-1/37 **S 92** II-1/37	**S 60, S 74:** two means to be provided, one being an engine room telegraph. **S 81, S 92:** two means to be provided, one being an engine room telegraph. Appropriate means to be provided to any other control position of the engines.
13	Engineers' alarm	**S 81** II-1/38 **S 92** II-1/38	**S 81, S 92:** to be operated from engine control room or platform and clearly audible throughout the accommodations.
14	Main source of electrical power and main lighting	**S 81** II-1/41 **S 92** II-1/41	**S 81, S 92:** one set sufficient for all main services.
15	Emergency source of electrical power (main supplies to be checked)	**S 60** II/26 **S 74** II-1/26 **S 81** II-1/43, 44 **S 92** II-1/43,44	**S 81, S 92:** capable of being readily started. If automatic start is provided, three consecutive starts to be performed and secondary means to be available in 30 min (manual start accepted). For supplies see annex 1.
16	Explosion proof lights in accumulator batteries stores, paint lockers, acetylene stores or similar spaces	**S81** II-1/45.10 **S92** II-1/45.10	See item below.
17	Precautions against shock and fire hazards are effective	**S 60** II/27 **S 74** II-1/27 **S 81** II-1/45 **S 92** II-1/45	**S 60, S 74:** check earth connection. Switchboards easily accessible (front and rear parts), non-conducting mats fitted. Hull return not allowed for tankers. Lights and wiring properly supported. Circuits protected against short circuit and overload. Batteries suitably housed in properly ventilated compartments. **S 81, S 92:** additionally, hull return not accepted for ships > 1600 GT. No electrical equipment admitted in dangerous areas unless certified as explosion proof.

No	Item	Ref.	Guidelines
18	Insulation mats around main and emergency switchboards	**S 60** II/27 **S 74** II-1/27(a)(ii) **S 81** II-1/45.2 **S 92** II-1/45.2	**S 60, S 74, S 81, S 92:** where necessary, non-conducting mats or gratings shall be provided at the front and rear of the switchboard.
19	Paint locker: all electrical equipment fitted inside is not-igniting and certified for safe usage in flammable gas mixtures	**S 81** II-1/45.10 **S 92** II-1/45.10	**S 81, S 92:** no electrical equipment shall be installed in paint lockers unless the Administration is satisfied that such equipment is: i. essential for operational purposes ii. of a type which will not ignite the mixture concerned iii. appropriate to the space concerned iv. appropriately certified for safe usage in the dusts, vapours or gases likely to be encountered. The following equipment may be installed in paint lockers provided that it is of an explosion-proof safe type: a. lighting fittings b. heating appliances c. ventilator motors. Switches, protective devices and motor control gear of the above services are preferably to be located outside the paint locker. **Ships built after 1.1.96:** the enclosed spaces giving access to the paint locker may be considered as non-hazardous provided that: 1. the door of the paint store is a gastight door with self closing devices without holding back arrangements (a watertight door may be considered as a gastight door) 2. the paint store is provided with independent natural ventilation system ventilated from a safe area; and 3. warning notices are fitted adjacent to the paint locker entrance stating that the locker contains flammable liquids.

No	Item	Ref.	Guidelines
20	Systems and alarms for unattended machinery spaces	**S 81** II-1/46-53 **S 92** II-1/46-53	**S 81, S 92:** alarms required for: fire in boiler air supply casings, scavenging air belts of main engines, engine bearings temperature (power ≥ 2250 kW or bore 〉 300 mm), high level in bilge wells, automatic shutdown of engines, automatic changeover of generators, power failure of alarm system. Alarms transferred to: machinery control position, engineers' public rooms and cabins, bridge.

SECTION J - FIRE FIGHTING

	General requirement for all fire-fighting appliances	**all ships** II-2/21	**All ships:** to be kept in good order and availability for immediate use at all times during the voyage.

J-1 - FIRE FIGHTING (MISCELLANEOUS)

No	Item	Ref.	Guidelines
1	Bulkheads and decks are in accordance with fire integrity requirements in the fire plan	**S 60** II/54 **S 74** II-2/51 **S 81** II-2/42-44, 46, 49, 50 **S 92** II-2/42-44, 46, 49, 50	**S 60, S 74: ships ≥ 4000 GT:** B-class bulkheads of corridors in accommodation. Steel stairtrunks. Kitchens, paint lockers, lockers, bosun stores adjacent to accommodations: steel bulkheads. **S 81, S 92:** one of three methods to be chosen: IC (B and C class divisions + smoke detection system in corridors, stairways, escape routes), IIC (sprinkler system with no restrictions on the type of bulkheads),IIIC (fire detection system with no restrictions on the type of bulkheads).

No	Item	Ref.	Guidelines
2	Means of escape: no obstructions, acceptable dead-end corridors, lighting, steps, handrails in good conditions	**S 60** II/68(b) **S 74** II-2/53 **S 81** II-2/45 **S 92** II-2/45	**S 60, S 74:** one means from accommodation and from spaces where crew normally work. Two from machinery spaces. **S 81, S 92:** two from accommodations, widely separated. Machinery spaces: two, widely separated, one being a continuous fire sheltered trunk from the lower level to an open deck. **S 81, S92:** dead end corridors (i.e. corridors provided with one escape route only): lengths of more than 7 m not acceptable
3	Fire detection and alarm system	**S 81** II-2/11, 52 **S 92** II-2/52	**S 81, S 92:** an automatic fire detection system is required in any machinery space where an automatic and remote control system in lieu of continuous manning is fitted.
4	Ventilation	**S 74** II-2/54(a) **S 81** II-2/48 **S 92** II-2/48	**S 74, S81, S 92:** main inlets and outlets can be closed from outside ventilated spaces. Power for ventilation can be shutdown from outside the ventilated spaces.
5	Openings (doors, ductwork, electrical wires, piping, etc.) do not impair the fire resistance of the bulkhead	**S 60** II/54 **S 74** II-2/51 **S 81** II-2/42-44, 46, 47, 49,50 **S 92** II-2/42-44, 46, 49, 50	**S 60, S 74, S 81, S 92:** not impairing the fire resistance of the bulkheads.
6	Automatic and manual fire doors	**S 81** II-2/47 **S 92** II-2/42-44, 46, 49, 50	**S 81, S 92:** doors in boundary bulkheads of cat. A machinery spaces reasonably gas-tight and self closing; not fitted with hold-back hooks unless fail safe type.

No	Item	Ref.	Guidelines
7	Remote means of closing fuel oil valves. In case shut-off valves are operated by air, check air cylinder pressure and pressure gauge	**S 60** II/69 **S 74** II-2/54 **S 81** II-2/15 **S 92** II-2/15	**S 60, S 74, S 81, S 92:** oil fuel transfer and similar pumps provided with remote controls for stopping outside machinery spaces. Oil fuel suction of storage, settling and service tanks above double bottom provided with valve capable of being closed outside the machinery spaces (S81: also lubricating oil).
8	Fire extinguishers (foam, CO_2, dry powder): –for each type, number indicated in the fire control plan available (see also item J-1.1) –condition of cylinders –validity of the extinguishing medium(*) (*) see Annex 2 to the Guidelines	**S 60** II/65 **S 74** II-2/52 **S 81** II-2/6 **S 92** II-2/6	**S 60, S 74: ships ≥ 1000 GT:** 5 in accommodation, 2+1 for each burner in boiler spaces, at least 1 each 1000 hp in machinery spaces (not less than 2, need not to exceed 6) + 1 foam (45 kg). **S 81, S 92:** to the satisfaction of the Administration. Ships 〈 1000 GT: not less than 5.
9	Portable foam applicator unit in boiler room and spaces containing internal combustion machinery: –check air-foam nozzle, portable tank of foam making liquid, spare tank, stowage container –test connection to fire main	**S 81** II-2/6.4, 7 **S 92** II-2/6.4, 7	**S 81, S 92:** there shall be in each boiler room at least one set of portable applicator units. A portable foam applicator unit shall consist of an air-foam nozzle of an inductor type capable of being connected to the fire main by a fire hose, together with a portable tank containing at least 20 l of foam-making liquid and one spare tank.
10	Foam type fire extinguisher of at least 135 l capacity in boiler room	**S 81** II-2/7.1.3 **S 92** II-2/7.1.3	**S 81, S 92:** there shall be not less than one approved foam-type extinguisher of at least 135 l capacity or equivalent in each boiler room.
11	Foam type fire extinguishers of at least 45 l capacity in spaces containing internal	**S 81** II-2/7.2 **S 92** II-2/7.2	**S 81, S 92:** each space containing internal combustion machinery shall be provided with a sufficient number of

No	Item	Ref.	Guidelines
	combustion machinery		foam type extinguishers (*) each of at least 45 l capacity or equivalent. (*) froth type according to S 74 and S 60
12	Fireman's outfits: –easily accessible –ready for use –stored in widely separated positions	**S 74** II-2/52(j) **S 81** II-2/17.4 **S 92** II-2/17.4	**S 74, S 81, S 92:** the fireman's outfits shall be so stored as to be easily accessible and ready for use and they shall be stored in widely separated positions.
13	Fireman's outfit: check availability of personal equipment	**all ships** II-2/17	**All ships:** fireman's outfit personal equipment shall consist of: –protective clothing of material to protect the skin from heat and burns and with water-resistant outer surface –boots and gloves –a rigid helmet –an electric safety lamp (hand lantern) –an axe
14	Fireman's outfit: check breathing apparatus	**all ships** II-2/17	**All ships:** a breathing apparatus may be either: 1. a smoke helmet or smoke mask which shall be provided with a suitable air pump and a length of air hose sufficient to reach from the open deck to any part of the holds or machinery spaces; or 2. a self contained compressed air-operated breathing apparatus, the volume of air contained in the cylinders of which shall be at least 1200 l, or other self-contained breathing apparatus which shall be capable of functioning for at least 30 minutes. A number of spare charges shall be provided. 3. a rigid helmet

No	Item	Ref.	Guidelines
15	Fireman's outfit: fireproof lifeline available	**all ships** II-2/17.2	**All ships:** for each breathing apparatus a fireproof lifeline of sufficient length and strength shall be provided
16	Paint locker: fire extinguishing arrangement	**all ships** II-2/18.7	Paint lockers shall be protected by an appropriate fire-extinguishing arrangement approved by the Administration. **1. paint locker deck area < 4 m2 :** portable fire extinguishers of CO_2 or dry chemical powder type to be provided as a minimum near the accesses of the paint locker. **2. paint locker deck area ≥ 4 m2 :** fixed fire extinguishing system to be provided to be operated from outside the paint locker. In case 2 above, the fixed system may use CO_2, dry chemical powder or pressure spray water. If a CO_2 system is provided, the quantity of CO_2 shall be sufficient to give a minimum volume of free gas equal to 40% of the gross volume of the paint locker. If a pressure spraying system is provided, the system may be fed by a connection from the ship's fire main. The number and arrangement of the nozzles shall be such as to ensure an effective distribution of water of at least 5 litres/m^2 per minute over the whole deck surface of the paint locker. If a dry chemical powder system is provided, the number and arrangement of the nozzles shall be such as to ensure an effective distribution of powder inside the paint locker with a ratio of at least 0.5 Kg/m^2.

No	Item	Ref.	Guidelines
17	Paint locker: structural fire protection	**S 60** II/54(e) **S 74** II-2/51(e) **S 81** II-2/44, 47, 58 **S 92** II-2/44,47,58	**S 60, S 74:** cargo ships ≥ 4,000 grt: paint stores when adjacent to accommodation spaces and emergency generator rooms, if any, shall be of steel or equivalent material. **S 81, S 92: cargo ships other than oil tankers:** paint stores are identified as high risk service spaces. The minimum fire integrity of bulkheads and decks shall be as prescribed in Tables 44.1 and 44.2. The fire resistance of doors shall, as far as practicable, be equivalent to that of the division in which they are fitted. **S 81, S 92: oil tankers:** paint stores are identified as high risk service spaces. The minimum fire integrity of bulkheads and decks shall be as prescribed in Tables 58.1 and 58.2. The fire resistance of doors shall, as far as practicable, be equivalent to that of the division in which they are fitted.
18	Arrangements for ro-ro cargo spaces	**S 81** II-2/53.2 **S 92** II-2/53.2	An approved automatic fire detection and fire alarm system to be provided. Spaces capable of being sealed to be protected by CO_2 or Halon. Other spaces: high expansion foam or water spray system. At least one portable fire extinguisher at each access to the cargo space. Three water fog applicators and one portable foam applicator unit for each cargo space, provided that the ship has at least two portable foam applicator units. Closed cargo spaces provided with ventilation system granting at least six air changes per hour. System to be

No	Item	Ref.	Guidelines
			entirely separated from the others: Capable of being controlled from the outside. Indicators of any loss of the required ventilating capacity to be provided on the navigating bridge. Arrangements for rapid shut-down and effective closure to be provided. Electrical equipment and wiring suitable for use in explosive mixtures except that above a height of 450 mm from the deck spark-preventing enclosed equipment shall be accepted provided that the ventilating system grants a continuous rate of at least ten air changes per hour.
19	Galleys exhaust ducts	**S 81** II-2/16.7 **S 92** II-2/48	**S 81, S 92:** exhaust ducts from galleys to be A class where they pass through accommodations or spaces containing combustible materials. Each exhaust duct to be fitted with grease trap, fire damper, fans shut-off within galley, fixed fire extinguishing system.
20	Requirements for tankers	**S 74** II-2/56, 57 **S 81** II-2/56-58 **S 92** II-2/56-58	**S 74:** A-60 separation between cargo pump room/cat. A machinery spaces and accommodation, service spaces, control stations. A-60 boundaries of superstructures where facing the cargo area (extended 3 m aft). No doors in front part of superstructures unless the boundaries of the space are A-60. fixed portlights (fitted with covers in the first tier). Self-closing doors in engine room casing without holdback hooks. Ventilation ducts provided with fire dampers where Cat. A machinery spaces and accommodation could be interconnected.

No	Item	Ref.	Guidelines
			S 81, S 92: as per S 74, except accommodation to be built according to IC method.
21	Requirements for tankers: ventilation in cargo areas	**S 74** II-2/58 **S 81** II-2/59 **S 92** II-2/59	**S 74:** openings of tanks far from sources of ignition. Inlets/outlets of superstructures not to permit gas penetration. Suction type ventilation for cargo pump room (20 changes per hour), outlets far from ignition sources. **S 81, S 92:** venting of cargo tanks to be distinct from air pipes of other compartments (could be incorporated with IGS). Outlets/inlets arranged to avoid risk of ignition and passage of flame. Venting provided with drainage to cargo tanks. Press-vac valves to be provided (at least 2 m from deck and 5 m from air intakes or sources of ignition). Provided with free flow system for vapours (at least 6 m from deck and 10 m from air intakes or sources of ignition) or high velocity valves (at least 2 m from deck and 10 m from air intakes or sources of ignition). Suction type ventilation for cargo pump room (20 changes per hour), outlets far from ignition sources; non-sparking type fans. **All oil tankers:** within 01/07/01 venting arrangements to be provided with a secondary means of allowing full relief of vapour, air or inert gas mixtures to prevent over pressure or under pressure in the system in case of failure of the primary means, such a second set of p/v valves, rupture disks, aliquid-filled p/v breaker or pressure sensors in each tank with a monitoring system and an

No	Item	Ref.	Guidelines
			alarm, activated by detection of over-pressure or under-pressure conditions within a tank

J-2 - FIRE MAIN SYSTEM

No	Item	Ref.	Guidelines
1	Main fire pumps and pressure gauges	**S 60** II/56, 65(b) **S 74** II-2/5, 52 **S 81** II-2/4 **S 92** II-2/4	**S 60, S 74:** total capacity of pumps in order to have water running at a speed of 2,7 m/s in a pipe having a diameter = 1,68 (L(B+D))0,5+25 (mm) (Total capacity needs not to exceed 180 t/h). Ships ≥ 1000 GT: at least two independent pumps (not in the same compartment, ships ≥ 2000 GT: a separated emergency pump to be provided). Bilge, ballast or sanitary pumps may be used as fire pumps. Each of the required fire pumps (apart from the emergency) shall have at least 80% of total capacity divided by the number of required pumps. **S 81, S 92:** additionally minimum delivery of pumps = 25 t/h. Emergency fire pump: minimum delivery 40% or 25 t/h. Diesel prime movers capable of being readily started. No direct access between EFP room and machinery spaces. In cargo ships with UMS, a remote starting of the fire pumps or permanent pressurisation is required. Isolating valves to be provided outside fire pumps spaces. All nozzles to be dual purpose type.
2	Simultaneous jets of water	**S 60** II/56© **S 74** II-2/5© **S 81** II-2/4.4	S 60, S 74: two pumps capable of delivering water to two nozzles (ships ≥ 6000 GT: 2,8 kg/sqcm, ships ≥ 1000

No	Item	Ref.	Guidelines
		S 92 II-2/4	GT: 2,6 kg/sqcm).
3	Check hydrants under pressure: –no leakage –fire hoses easily coupled –condition of valves	**S 60** II/56, 65(b) **S 74** II-2/5, 52 **S 81** II-2/4 **S 92** II-2/4	**S 60: ships ≥ 1000 GT:** two jets in each part of the ship. One hose+nozzle each 30 m of ship's length (at least 5), excluding machinery spaces. Unless one hose+nozzle is provided for each hydrant they must be interchangeable.
4	Check nozzles: –no leakage (for nozzles under pressure) –dual type where required	**S 74** II-2/5(g) **S 81** II-2/4.8 **S 92** II-2/4.8	**S 74:** dual purpose type in machinery spaces or in similar spaces where the risk of spillage of oil exists **S 81, S 92:** all nozzles shall be dual purpose type (i.e. spray/jet type) incorporating a shut off
5	International shore connection	**S 60** II/56(h) **S 74** II-2/52(d) **S 81** II-2/19 **S 92** II-2/19	**S 60, S 74: ships ≥ 1000 GT:** one ISC to be provided, available for being used on either side of the ship. Outside/inner diameters = 178/64 mm. Bolt circle diameter = 132 mm. Four slotted holes, 19 mm in diameter. Thickness = 14,5 mm. Four bolts, 16 mm in diameter, 50 mm in length. **S 81, S 92: ships ≥ 500 GT:** as above.

J-3 - CO_2 SYSTEM(S)

S 60 II/58	S 74 II-2/7	S 81 II-2/5
BOILERS (GT ≥ 1000) OIL FUEL UNITS (GT ≥ 1000) ENGINES (GT ≥ 1000, P ≥ 750 kW) DRY CARGO (GT ≥ 2000) **(1)** CARGO TANKS (GT ≥ 2000) **(2,3) (*)** RO-RO SPACES	BOILERS (GT ≥ 1000) OIL FUEL UNITS (GT ≥ 1000) ENGINES (GT ≥ 1000, P ≥ 750 KW) DRY CARGO (GT ≥ 2000) **(1)** RO-RO SPACES	BOILERS (GT ≥ 1000) OIL FUEL UNITS (GT ≥ 1000) ENGINES (GT ≥ 1000, P ≥ 375 KW) DRY CARGO (GT ≥ 2000) **(1)** RO-RO SPACES DANGEROUS GOODS CARGO PUMP ROOMS
(*) see the note after item J-10 of these Guidelines		
CO_2 (0,56 m^3/kg) available for 30% of the gross volume of the biggest cargo compartment served or 40% machinery spaces plus casing (partial) / 35% machinery spaces plus casing (the greatest). Each pipe provided with a valve operable from the deck clearly marked with the name of the compartment served. Big holds provided with fwd and aft pipes. Shutdown for ventilators and closing appliances operable from outside the compartment served. Sound alarm provided before discharge.	CO_2 (0,56 m^3/kg) available for 30% of the gross volume of the biggest cargo compartment served or 40% machinery spaces plus casing (partial) / 35% machinery spaces plus casing (the greatest), free air in air containers to be taken into account. Each pipe provided with a valve operable from the deck clearly marked with the name of the compartment served. Big holds provided with fwd and aft pipes. Shutdown for ventilators and closing appliances operable from outside the compartment served. Sound alarm provided before discharge. Bottle storage room to be in safe and accessible position and effectively ventilated, separated from compartment served and with gastight boundaries.	CO_2 (0,56 m^3/kg) available for 30% of the gross volume of the biggest cargo compartment served or 40% machinery spaces plus casing (partial) / 35% machinery spaces plus casing (the greatest), free air in air containers to be taken into account. Each pipe provided with a valve operable from the deck clearly marked with the name of the compartment served (clear instructions to be posted). Big holds provided with fwd and aft pipes. Shutdown for ventilators and closing appliances operable from outside the compartment served. Automatic sound alarm provided before discharge (suitably delayed). Bottle storage room to be in safe and accessible position and effectively ventilated, separated from compartment served and with gastight boundaries and outwards opening door. Means provided to crew for checking the content of bottles.

J-4 - HALON SYSTEM(S)

S 81 II-2/5
BOILERS (GT ≥ 1000) OIL FUEL UNITS (GT ≥ 1000) ENGINES (GT ≥ 1000, P ≥ 375 KW) DRY CARGO (GT ≥ 2000) **(1)** RO-RO SPACES DANGEROUS GOODS CARGO PUMP ROOMS
Halon 1301 available for 5%-7% (4,25%~7%) of the gross volume of the biggest cargo vehicles (machinery) compartment served. Each pipe provided with a valve operable from the deck clearly marked with the name of the compartment served (clear instructions to be posted). Big holds provided with fwd and aft pipes. Automatic shutdown for ventilators and manually operated closing appliances operable from outside the compartment served. Automatic sound alarm provided before discharge (suitably delayed). Bottle storage room to be in safe and accessible position and effectively ventilated, separated from compartment served and with gastight boundaries and outwards opening door. Means provided to crew for checking the content of bottles.

J-5 - FOAM SYSTEM(S)

S 60 II/60	S 74 II-2/10, 9	S 81 II-2/9, 8
BOILERS (GT ≥ 1000) OIL FUEL UNITS (GT ≥ 1000) ENGINES (GT ≥ 1000, P ≥ 750 kW) CARGO TANKS (GT ≥ 2000) **(2,3)** CARGO DECK (GT ≥ 2000) **(2,3)**	BOILERS (GT ≥ 1000) OIL FUEL UNITS (GT ≥ 1000) ENGINES (GT ≥ 1000, P ≥ 750 KW) CARGO TANKS (GT ≥ 2000) **(2,3)**	BOILERS (GT ≥ 1000) OIL FUEL UNITS (GT ≥ 1000) ENGINES (GT ≥ 1000, P ≥ 375 KW) RO-RO SPACES CARGO PUMP ROOMS
Foam available for 15 cm height on the largest compartment floor. System operable from outside the compartment served.	High expansion: foam available for protecting the largest compartment at a rate of at least 1 m in depth per minute. Foam available	High expansion: foam available for protecting the largest compartment at a rate of at least 1 m in depth per minute. Foam available

S 60 II/60	S 74 II-2/10, 9	S 81 II-2/9, 8
	for five times the biggest volume served. Expansion ratio not to exceed 1000:1. Froth generators, sources of power supply, froth forming liquid and means of control grouped and readily accessible from outside the compartment served. Low expansion: foam available for 15 cm height on the largest compartment floor. System operable from outside the compartment served. Expansion ratio not to exceed 12:1.	for five times the biggest volume served. Expansion ratio not to exceed 1000:1. Froth generators, sources of power supply, froth forming liquid and means of control grouped and readily accessible from outside the compartment served. Low expansion: foam available for 15 cm height on the largest compartment floor. System operable from outside the compartment served. Expansion ratio not to exceed 12:1.

J-6 - SPRINKLER SYSTEM

S 81 II-2/12
ACCOMMODATION **(IIC)**
Each section of sprinklers shall include fail-safe means for giving a visual and audible alarms signal automatically at one indicating unit on the navigating bridge whenever any sprinkler comes into operation. The unit shall indicate the section where fire has occurred and audible and visible alarms shall be repeated in another position to ensure immediate reception by crew. A plan of the system and areas covered to be provided near the unit. Unit provided with test arrangement. Each section capable of being isolated by one stop valve only, provided with pressure gauge, readily accessible by authorised persons and clearly indicated, test valve to be provided. Approved sprinkler heads o provide at least 5 l/m^2 per minute. A pressure tank connected to air system and fresh water for replenishing to keep system under pressure and an independent pump, automatically starting for pressure decrease and fitted with test valve to be provided remote from any category A machinery space. Pump, alarms and detection powered by two supplies. System connected via a non-return valve to the fire main.x

J-7 - WATER SPRAY SYSTEM(S)

S 60 II/51	S 74 II-2/11	S 81 II-2/10
BOILERS (GT ≥ 1000) OIL FUEL UNITS (GT ≥ 1000) ENGINES (GT ≥ 1000, P ≥ 750 kW)	BOILERS (GT ≥ 1000) OIL FUEL UNITS (GT ≥ 1000) ENGINES (GT ≥ 1000, P ≥ 750 KW) CARGO PUMP ROOMS **(2)**	BOILERS (GT ≥ 1000) OIL FUEL UNITS (GT ≥ 1000) ENGINES (GT ≥ 1000, P ≥ 375 KW) RO-RO SPACES CARGO PUMP ROOMS
Type approved nozzles. The system can be divided in some sections. Pump to start automatically for pressure decrease, to deliver sufficient water to all sections, to be outside the compartment served.	Type approved nozzles providing at least 5 litres per square metre per minute. The system can be divided in some sections. Pump to start automatically for pressure decrease, to deliver sufficient water to all sections, to be outside the compartment served. Pump to be driven by independent engine or supplied by emergency generator.	Type approved nozzles providing at least 5 litres per square metre per minute. The system can be divided in some sections. Pump to start automatically for pressure decrease, to deliver sufficient water to all sections, to be outside the compartment served. Pump to be driven by independent engine or supplied by emergency generator.

J-8 - STEAM SYSTEM(S)

S 60 II/58	S 74
DRY CARGO (GT ≥ 2000) **(1)** CARGO TANKS (GT ≥ 2000) **(2,3)** RO-RO SPACES	RO-RO SPACES
Boilers for production of fire-fighting steam to have at least 1 kg/h per 0,75 m^3 gross volume of the biggest compartment served. Each pipe provided with a valve operable from the deck clearly marked with the name of the compartment served. Big holds provided with fwd and aft pipes going to the lowest part of the hold. Shutdown for ventilators and closing appliances operable from outside the compartment served.	Boilers for production of fire-fighting steam to have at least 1 kg/h per 0,75 m^3 gross volume of the biggest compartment served. Each pipe provided with a valve operable from the deck clearly marked with the name of the compartment served. Big holds provided with fwd and aft pipes going to the lowest part of the hold. Shutdown for ventilators and closing appliances operable from outside the compartment served.

J-9 - DECK FROTH SYSTEM

S 60	S 74 II-2/61	S 81 II-2/61
CARGO DECK (GT ≥ 2000) **(2,3)**	CARGO DECK (GT ≥ 2000) **(2,3)**	CARGO DECK
	Main control station outside of cargo tank area, adjacent to accommodation, readily accessible and operable. Rate of supply not less than 0,6 litres per minute per square metre (ship's breadth x cargo tanks total length) or 6 litres per minute per square metre (horizontal area of the greatest cargo tank) (the greater) for at least 20 minutes. Expansion ratio not to exceed 150:1. Each monitor to provide at least 50% required froth rate. A monitor and a hose connection for a froth applicator provided each side accommodation front facing cargo area. Valves provided in both the froth main and the fire main, fwd each monitor. Operation of froth system at nominal output to allow contemporary use of fire main with two jets of water.	Main control station outside of cargo tank area, adjacent to accommodation, readily accessible and operable. Rate of supply not less than 0,6 litres per minute per square metre (ship's breadth x cargo tanks total length) or 6 litres per minute per square metre (horizontal area of the greatest cargo tank) or 6 litres per minute per square metre (area protected by largest monitor, not less than 1250 l/minute) (the greater) for at least 20 minutes (30 if no IGS provided). Expansion ratio not to exceed 150:1. Each monitor to provide at least 50% required froth rate. A monitor and a hose connection for a froth applicator provided each side accommodation front facing cargo area. At least 4 applicators to be provided, capacity 400 l and throw 15 m. Valves provided in both the froth main and the fire main, fwd each monitor. Operation of froth system at nominal output to allow contemporary use of fire main with two jets of water.

J-10 - INERT GAS SYSTEM

S 74 (S 81 II-2/62.20)	S 81 II-2/62
CARGO TANKS **(3)**	CARGO TANKS **(3)**
Two fans to be provided, together capable of delivering at least 125% the maximum rated capacity of the cargo pumps. Scrubber provided to cool gas and remove solid and sulphur products. Valves with position indicators to be provided between boiler and scrubber. A gas regulating valve to be provided in the supply main. Following to be provided: non-return valve with positive means of closing and deck water seal supplied by two pumps for protecting machinery spaces, P/V valves for protecting cargo tanks, P/V breaker on the supply main, stop valve on inert gas supply for each tank, continuous recording of oxygen content and inert gas pressure (in cargo control room), portable oxygen analyser, inert gas temperature and pressure indicators (in cargo control room), means for the zero and span calibration of fixed and portable oxygen analysers, alarms for: high oxygen content (exceeding 8%), low gas pressure (less than 100 mm water gauge) , low water level in deck water seal, high gas temperature (*), high gas pressure, low pressure or flow of water to the scrubber (*), high water level in the scrubber (*), failure of inert gas blowers, failure of power supply to control system of gas regulating valve. Automatic shut-down of blowers and gas regulating valve provided for (*) alarms. Gas regulating valve to shutdown automatically for blowers failure.	Two fans to be provided, together capable of delivering at least 125% the maximum rated capacity of the cargo pumps. Scrubber provided to cool gas and remove solid and sulphur products. Oxygen content in inert gas not to exceed 5%. Valves with position indicators to be provided between boiler and scrubber. A gas regulating valve to be provided in the supply main. Following to be provided: non-return valve with positive means of closing and deck water seal supplied by two pumps for protecting machinery spaces, P/V valves for protecting cargo tanks, P/V breaker on the supply main, stop valve on inert gas supply for each tank, continuous recording of oxygen content and inert gas pressure (in cargo control room), portable oxygen analyser, inert gas temperature and pressure indicators (in cargo control room), means for the zero and span calibration of fixed and portable oxygen analysers, alarms for: high oxygen content (exceeding 8%), low gas pressure (less than 100 mm water gauge) , low water level in deck water seal, high gas temperature (*), high gas pressure, low pressure or flow of water to the scrubber (*), high water level in the scrubber (*), failure of inert gas blowers, failure of power supply to control system of gas regulating valve. Automatic shut-down of blowers and gas regulating valve provided for (*) alarms. Gas regulating valve to shutdown automatically for blowers failure.

(1) The Flag Administration may grant exemption under the following conditions (S 48, S 60: A or B or C, S 74/78: A+B or C, S 81: A+B):

A. cargo spaces provided with steel hatch covers and effective means of closing all ventilators and other openings leading to the holds,

B. ship constructed and intended solely for carrying such cargoes as ore, coal, grain, unseasoned timber, and non-combustible cargoes or cargoes which, in the opinion of the Administration, constitute a low fire risk,

C. ship engaged on voyages of short duration (in the opinion of the Administration).

(2) Foam system to be provided for protecting either cargo tanks or cargo deck area.

(3) The following ships are to be provided with an Inert Gas System:

i. all tankers, built on or after 1.6.80, DWT $\geq$ 20000

ii. crude oil tankers, built before 1.6.80, 20000 $\leq$ DWT $<$ 40000, delivery of each washing machine $\geq$ 60 m^3/h (*)

iii. crude oil tankers, built before 1.6.80, DWT $\geq$ 40000

iv. product carriers, built before 1.6.80, 20000 $\leq$ DWT $<$ 40000, delivery of each washing machine $\geq$ 60 m^3/h (**)

v. product carriers, built before 1.6.80, DWT $\geq$ 40000

(*) if delivery of each washing machine $<$ 60 m^3/h, exemption may be granted

(**) if delivery of each washing machine $<$ 60 m^3/h, exemption is automatic

SECTION K - LIFESAVING APPLIANCES

	General requirement for lifesaving appliances	**all ships** III/19.2	**All ships:** to be kept in good order and availability for immediate use at all times during the voyage.

K-1 - LIFEBOATS

No	Item	Ref.	Guidelines
1	Required number	**S 60** III/8, 35 **S 74** III/8, 35 **S 92** III/26	**S 60, S 74:** one open lifeboat on each side for the total number of persons on board. Ships ≥ 1600 GT: at least one to be a motor lifeboat (tankers: one for each side). **S 92:** one totally enclosed lifeboat on each side for the total number of persons on board or one free-fall totally enclosed lifeboat after. Ships having length 〈 85 m, other than oil tankers, chemical tankers and gas carriers may have only liferafts. Lifeboats for chemical tankers and gas carriers carrying toxic cargoes shall be provided also with a self- contained air support system. Lifeboats for oil tankers, chemical tankers and gas carriers carrying cargoes with flashpoint ≤ 60°C shall also be fire-protected. Additionally, a rescue boat shall be provided.
2	Inventory	**S 60** III/11, 12 **S 74** III/11, 12 **S 92** III/41.8, 47	See annexes 3, 4, 5.
3	Lifeboats properly marked	**S 60** III/20 **S 74** III/20 **S 92** III/41.9, 47.1	**S 60, S 74:** dimensions, number of persons, ship's name. **S 92:** lifeboats and rescue boat: dimensions, number of person, ship's name, port of registry on each side; ship's identification and number of the lifeboat on top.

No	Item	Ref.	Guidelines
4	Retro-reflective tapes	**all ships** III/30.2.7	**all ships:** refer to the Recommendation on the Use and Fitting of Retro-Reflective Materials on Life Saving Appliances adopted by IMO Res. A.658(16) made mandatory under SOLAS
5	Embarkation arrangement	**S 60** III/19 **S 74** III/19 **S 92** III/11, 15	**S 60, S 74, S 92:** one embarkation ladder for each davit (S 92: not required for free-fall lifeboats). Embarkation and lowering areas properly lighted (also from emergency source), means to stop water from ship to lifeboats.
6	Winches, davits, cables, blocks and sheaves	**S 60** III/36 **S 74** III/36 **S 92** III/15, 48	**S 60, S 74: tankers ≥ 1600 GT:** gravity type. Other ships: luffing type accepted only where lifeboat's weight 〈 2300 kg. Limit switches if mechanical hoisting is provided. **S 92:** gravity or free-fall type, provided with launching and recovery arrangements. Limit switches if mechanical hoisting is provided. Launching system from survival craft provided.
7	Operating instructions posted	**all ships** III/9	**All ships:** to be posted near the launching appliances where lighted by the emergency lighting, indicating procedures for launching, adopting IMO blue symbols.
8	Thorough examination and dynamic test	**all ships** III/20.11.1	**All ships:** launching appliances to be serviced at recommended intervals in accordance with onboard maintenance instructions and subject at intervals not exceeding 5 years to a thorough examination(to be carried out with the same philosophy adopted for cargo lifting appliances) and a dynamic test of the winch brake

No	Item	Ref.	Guidelines
9	On-load release gear	**all ships** III/20.11.2	**All ships:** lifeboats on-load release gear to be serviced at recommended intervals in accordance with onboard maintenance instructions and to a thorough examination and test during the surveys of the Safequ/Safpas Certificates by properly trained personnel; they shall be also operationally tested under an over load whenever the release gear is overhauled (at least once every 5 years)
10	Falls used in launching	**all ships** III/20.4	**All ships:** to be turned end for end every 30 months and renewed when necessary but not later than 5 years or alternatively to be renewed not later than 4 years without end-for-ending

K-2 - RESCUE BOATS

No	Item	Ref.	Guidelines
1	Rescue boats	**S 92** III/26.2,47	**S 92:** cargo ships shall carry at least one rescue boat complying with the requirements of Reg. III/47. A lifeboat may be accepted as a rescue boat provided that it also complies with the requirements for a rescue boat.

K-3 - LIFERAFTS

No	Item	Ref.	Guidelines
1	Required number	**S 92** III/26.1 except all ships built before 01.07.86: **S 92** III/26.3	**New ships:** one or more rigid or inflatable liferafts for the total number of persons on board where they can be readily transferred on either side of the ship, otherwise total number required

No	Item	Ref.	Guidelines
			on each side. Where a free-fall lifeboat is provided, liferafts for the total number of persons on board to be provided on each side of the ship and at least one side shall be served by launching appliances. Ships having length 〈 85 m, other than oil tankers, chemical tankers and gas carriers provided only with liferafts shall have the above mentioned liferafts provided that they are sufficient for the total number of persons on each side even when one is lost or damaged (the transferring criteria can be used). If they are not readily transferrable, they must be sufficient for 150% the total persons on each side. Additionally, where survival craft are stowed more than 100 m from the stem or stern, a liferaft stowed as far as possible shall be provided. **Existing ships:** one or more rigid or inflatable liferafts for the total number of persons on board. Additionally, where survival craft are stowed more than 100 m from the stem or stern, a liferaft stowed as far as possible shall be provided.
2	Stowage	**S 92** III/26.1.4, 29 except all ships built before 01.07.86: **S 92** III/26.3.2, 29	**New ships:** any liferaft to be stowed being connected to the ship through a float-free arrangement (i.e.: a hydrostatic release unit) with its painter permanently attached to the ship through a weak link. The additional "more than 100 m" liferaft may be securely fastened (not float-free) so as to permit manual release and need not launching appliances.

No	Item	Ref.	Guidelines
			Existing ships: any liferaft to be stowed being connected to the ship through a float-free arrangement (i.e.: a hydrostatic release unit) with its painter permanently attached to the ship through a weak link. The additional "more than 100 m" liferaft may be securely fastened (not float-free) so as to permit manual release.
3	Inflatable liferafts servicing	**all ships** III/19.8.1	**All ships:** every 12 months
4	Hydrostatic release units servicing	**all ships** III/19.9.1	**All ships:** every 12 months
5	Marking	**S 60** III/20 **S 74** III/20 **S 92** III/39.7.3,40.7 **all ships** III/38.6.3.5	**S 60, S 74:** number of persons, serial number, maker (inflatable), number of persons, ship's name, port of registry (rigid). **S 92:** inflatable: maker's name, serial number, name of approving authority, SOLAS, type of emergency pack enclosed, date of last servicing, length of painter, maximum permitted height of stowage, launching instructions. Rigid liferafts: as above (apart from date of servicing) + ship's name and port of registry, number of persons. **All ships:** hydrostatic release unit permanently marked with type and serial number.
6	Embarkation arrangement	**S 60** III/19 **S 74** III/19 **S 92** III/11	**S 60, S 74:** one embarkation ladder for each davit. Embarkation and lowering areas properly lighted, means to stop water from ship to lifeboats.
7	Winches, davits, cables, lifting hooks (if applicable)	**S 92** III/48	**S 92:** gravity launching type. Launching system from survival craft provided.

K-4 - LIFEBUOYS

No	Item	Ref.	Guidelines
1	Required number	**S 60** III/37 **S 74** III/37 **S 92** III/27	**S 60, S 74:** at least 8, half provided with automatic light (tankers: battery powered) two of them also with smoke signal, one for each side provided with 27,5 m lifeline. Immediately available and marked with ship's name and port of registry. **S 92:** 8 ($L < 100$ m), 10 ($L < 150$ m), 12 ($L < 200$ m), 14 ($L \geq 200$ m), half provided with self-igniting light (tankers: battery powered) two of them also with smoke signal and quickly releasable from the bridge, one for each side provided with a lifeline of 30 m or twice the stowage height (the greater). Immediately available and marked with ship's name and port of registry.
2	Retro-reflective material	**all ships** III/30.2.7	**All ships:** all lifesaving appliances to be fitted with retro-reflective material according to IMO recommendations.

K-5 - LIFEJACKETS

No	Item	Ref.	Guidelines
1	Required number	**S 60** III/22 **S 74** III/22 **S 92** III/7.2	**S 60, S 74:** one for each person, also inflatable (tankers: not accepted) and provided with whistle. Immediately available, position clearly marked. **S 92:** one for each person + suitable number for persons on watch and for use at remotely located survival craft stations, also inflatable and provided with whistle. Immediately available, position clearly marked.

No	Item	Ref.	Guidelines
2	Inflatable lifejackets servicing	**all ships** III/19.8.1	**All ships:** every 12 months
3	Lights	**all ships** III/27.2	**All ships:** each lifejacket to be provided with a light.
4	Retro-reflective material	**all ships** III/30.2.7	**All ships:** all lifesaving appliances to be fitted with retro-reflective material according to IMO recommendations.

K-6 - IMMERSION SUITS AND THERMAL PROTECTIVE AIDS

No	Item	Ref.	Guidelines
1	Required number	**all ships** III/27.3 **S 92** III/7.3	**All ships:** three immersion suits for each lifeboat and one thermal protective aid for each person not provided with an immersion suit where lifeboats are open or partially enclosed type. Ships provided only with liferafts to have an immersion suit for each person on board, unless liferafts are davit-launched. **S 92:** one immersion suit for every person assigned to crew the rescue boat. This requirement can be fulfilled by the above.
	Retro-reflective material	**all ships** III/30.2.7	**All ships:** all lifesaving appliances to be fitted with retro-reflective material according to IMO recommendations.

K-7 – LIVESAVING AND SAFETY EQUIPMENT (MISCELLANEOUS)

No	Item	Ref.	Guidelines
1	Stowage locations	**S 92** III/20.10	**All ships:** containers, brackets, racks and other similar stowage locations for life-saving equipment shall be marked with symbols in accordance with IMO Res. A.760(18), indicating the devices stowed in that location and their number, if more than one device is stowed.
2	Distress signals	**S 60** III/24 **S 74** III/24 **S 92** III/6.3	**S 60, S 74:** number to the satisfaction of the Administration (12 parachute to be included).
3	Two-way VHF radiotelephone apparatus	**all ships** III/6.2.1	**All ships: 300 ≤ GT < 500:** two two-way VHF radiotelephone apparatus to be provided; GT ≥ 500: three two-way VHF radiotelephone apparatus to be provided. Apparatus installed before 01.02.92 can be accepted until 01.02.99 if compatible to the new standards.
4	Radar transponders	**all ships** III/6.2.2	**All ships: 300 ≤ GT < 500:** one radar transponder to be provided; **GT ≥ 500:** one radar transponder for each side to be provided. They are to be stowed in way they can be rapidly placed into any survival craft apart from liferafts required by III/26.1.4. Alternatively, one radar transponder may be placed in any survival craft (liferafts included).
5	Line-throwing appliances	**S 60** III/23 **S 74** III/23 **S 92** III/17, 49	**S 60, S 74, S 92:** one with 4 projectiles and 4 lines.
6	On-board communication and alarm system	**S 92** III/4.1	**S 92 (01.07.86):** an emergency means comprising either fixed or

No	Item	Ref.	Guidelines
			portable equipment or both shall be provided for two-way communications between emergency control stations, muster and embarkation stations and strategic positions on board.
7	General alarm system	**S 92** III/4.2, 50	**S 92 (01.07.86):** provided to be used for summoning crew to muster stations and to initiate the actions included in the muster list. Capable of sounding the general alarm signal on the ship's whistle and on an electrically operated warning system. Operable from bridge and other strategic positions (not ship's whistle). Audible throughout accommodations and crew areas. Powered also by emergency source. To be supplemented by either a public address system or other suitable means of communication.

SECTION L - POLLUTION PREVENTION

L-1 - ANNEX I: OIL

No	Item	Ref.	Guidelines
1	Oily-water separator	**M** I/9, 16	**All ships delivered before 6.7.93:** may discharge outside special areas into the sea through a 15 ppm oily-water separating equipment (100 ppm accepted until 6.7.98). Ships ≥ 10000 GT shall be also provided with an oil discharge monitoring and control system unless they are provided with a 15 ppm oily-water separating equipment. Ships delivered after 6.7.93

No	Item	Ref.	Guidelines
			may discharge outside special areas into the sea through a 15 ppm oily-water separating equipment. Ships ≥ 10000 GT shall be also provided with a 15 ppm alarm and automatic stop of discharge. Ships trading in special areas can retain oily waters on board in an holding tank of adequate volume. All oily water to be discharged to reception facilities, quantity, time and port of discharge to be recorded in the Oil Record Book. Discharge of oily waters in special areas are permitted through 15 ppm separators provided with 15 ppm automatic stopping device.
2	Segregation of oil fuel and water ballast systems	**M I**/14	New oil tankers ≥ 150 GT and other new ships ≥ 4000 GT shall not carry ballast water in any oil fuel tank. Ships ≥ 400 GT, built on or after 1.7.82 shall not carry oil fuel in a forepeak or in a compartment forward the collision bulkhead.
3	Sludge tank arrangements	**M I**/17, 19	Compulsory for all ships ≥ 400 GT. To be connected only to two standard discharge connections (outside diameter = 215 mm, inner diameter according to pipe outside diameter, bolt circle diameter = 183 mm, 6 slotted holes 22 mm diameter, flange thickness = 20 mm, 6 bolts 20 mm diameter suitable length).
4	Oil Discharge Monitoring and Control System (ODMS)	**M I**/15, 16	Compulsory for all oil tankers ≥ 150 GT, residues to be transferred to slop tank(s) (at least two for new oil tankers ≥ 70000 t DWT). An oil discharge

No	Item	Ref.	Guidelines
			monitoring and control system to be provided. Effective oil/water interface detectors to be provided. Some exemptions granted for restricted service.
5	Segregated ballast tanks system	**M I**/13	Compulsory for all new crude oil tankers ≥ 20000 t DWT, all new product carriers ≥ 30000 t DWT, all existing crude oil tankers ≥ 40000 t DWT unless provide with COW system
6	Crude Oil Washing (COW) system	**M I**/13	Compulsory for all new crude oil tankers ≥ 20000 t DWT
7	Check discharge piping for dirty or oilcontaminated water and part flow system	**M I**/18	One manifold for connection to reception facilities compulsory for all oil tankers on the open deck on both sides of the ship. Pipelines for discharge of ballast or oily waters from cargo area to be led to open deck or to the ship's side above waterline. Stoppage from open deck where discharge is visible to be provided in new oil tankers.

L-2 - ANNEX II: CHEMICAL

No	Item	Ref.	Guidelines
1	Chemical tankers Oil tankers with NLS/IPPC certificate Gas tankers with NLS/IPPC certificate	M II/5A, Standards for procedures and arrangements	**New:** ships constructed on or after 1.7.86, Ex: other ships A, B, C, D: category A, B, C, D substances Every ship to be provided with pumping and piping arrangements such as to limit the residue in pipes and suction well to a quantity of 0,1 m^3 (New-B), 0,3 m^3 (New-C, EX-B), 0,9 m^3 (Ex-C). Ex need not to comply if prewash procedure is adopted and subsequent washing or

No	Item	Ref.	Guidelines
			ballast water is discharged to reception facilities. Some substances require a cargo heating system. The underwater discharge of noxious substances is to be fitted in the cargo area near the turn of bilge. Slop tanks are not compulsory but they are needed for some washing procedures. Cargo tanks may be used for this purpose. Residues of cargo may be removed by means of ventilation provided that precautions are taken against ignition. All a.m. procedures to be in accordance with the ship's P&A Manual approved by the Administration.

SECTION M - RADIO EQUIPMENT

Guidelines : T = transmitter; R = receiver

GMDSS SHIPS

■ : compulsory item,
▲x: optional compulsory item,
: identifies a set of items which may be an alternative to the set identified by .

No	Guidelines	Ref.	A1	A2	A3	A4
1	Electrical lighting, independent of the main and emergency sources of the ship, for radio controls. Call sign, ship station identity and other required codes to be clearly marked. Controls of the VHF RT safety channels shall be immediately available on the bridge convenient to the conning position. Adequate tools and spares for maintenance shall be provided.	**S92** IV/4, 6	■	■	■	■
2	**Radio personnel:** all ships shall carry personnel qualified for distress and safety radiocommunication purposes to the satisfaction of the Administration. They shall be holders of certificates specified in the Radio Regulations.	**S92** IV/16	■	■	■	■

No	Guidelines	Ref.	A1	A2	A3	A4
3	**Radio logs:** a record shall be kept of all incidents connected to the radiocommunication service relevant to safety.	**S92** IV/17	■	■	■	■
4	**Sources of energy:** there shall be available at all times a supply of energy sufficient to operate the radio installations and to charge any batteries used as part of a reserve source. A reserve source shall be available in the event of failure of the ship's main and emergency sources and shall be independent of the propelling power of the ship. Where batteries are used: a means for automatically charging to minimum capacity within 10 h shall be provided, the capacity shall be checked once a year (not at sea) by fully discharging using normal current and period (10 h). The reserve source shall supply: VHF (7.1.1), MF radio installation (9.1.1), MF/HF radio installation (10.2.1 or 11.1), INMARSAT ship earth station (10.1.1), all radio installations capable of being connected to the reserve source, electrical lighting, devices for receiving input of information. The period of supply shall be 1 h for ships built before 1.2.95, if they have an emergency source in full compliance with S92, otherwise shall be 6 h. It shall be 1 h for ships built on or after 1.2.95.	**S92** IV/13	■	■	■	■
5	**Maintenance requirements:** availability of the functional requirements shall be ensured by using one (A1, A2) or two (A3, A4) of the following choices: a) duplication of equipment, b) shore-based maintenance, c) at-sea electronic maintenance.	**S92** IV/15	■	■	■	■
6	VHF T+R operating on DSC frequency 156.525 MHz (channel 70) and on RTP frequencies 156.300 MHz (channel 6), 156.650 MHz (channel 13), 156.800 MHz (channel 16). Incorporated VHF DSC watch receiver on channel 70 (may be separated). A1: capable also of transmitting and receiving general RTP radiocommunications (may be separated).	**S92** IV/7.1.1 **S92** IV/7.1.2 **S92** IV/8.2	■	■	■	■

No	Guidelines	Ref.	A1	A2	A3	A4
7	9 Ghz radar transponder stowed for easy utilisation (may be one of those used for survival craft if readily available).	**S92** IV/7.1.3	■	■	■	■
8	NAVTEX receiver for international service broadcasts	**S92** IV/7.1.4	■	■	■	■
9	INMARSAT radio facility for reception of maritime safety information on enhanced group calling system if ship engaged where NAVTEX coverage is not provided.	**S92** IV/7.1.5	▲1	▲1	▲1	▲1
10	HF maritime system information receiver using direct-printing telegraphy if ship engaged only in areas where such service is provided.	**S92** IV/7.1.5	▲1	▲1	▲1	▲1
11	Satellite emergency position-indicating radio beacon (satellite EPIRB) installed in easily accessible position, ready to be manually released and activated, capable of floating free and being automatically activated when afloat and capable of transmitting a distress alert in the 406 MHz band. A1 3: may be used as a secondary means of ship-to-shore alert from bridge operating on 406 MHz or a separate item should be provided. A2 4, A3 6 7, A4 : may be used as a secondary means of ship-to-shore alert from the bridge by a radio service (A2: other than MF, A3 7 , A4: other than HF) operating on 406 MHz or a separate item should be provided.	**S92** IV/7.1.6 **S92** IV/8.1.2 **S92** IV/9.1.3.1 **S92** IV/10.1.4.1 **S92** IV/10.2.3.1 **S92** IV/11				
12	Satellite emergency position-indicating radio beacon (satellite EPIRB) installed in easily accessible position, ready to be manually released and activated, capable of floating free and being automatically activated when afloat and capable of transmitting a distress alert through the INMARSAT satellite in the 1.6 Ghz band, if ship engaged on voyages where only this coverage is provided. A1 3: may be used as a secondary means of ship-to-shore alert from bridge operating through the INMARSAT service or a separate item should be	**S92** IV/7.1.6 **S92** IV/8.1.5.2 **S92** IV/10.1.4.3 **S92** IV/10.2.3.2.1 **S92** IV/9.1.3.3.2 **S92** IV/10.2.3.2.2 **S92** IV/11				

No	Guidelines	Ref.	A1	A2	A3	A4
	provided. A2 4, A3 6 7 , A4 8: may be used as a secondary means of ship-to-shore alert from the bridge by a radio service (A2: other than MF, A3, A4: other than HF) operating through the INMARSAT service or a separate item should be provided.					
13	Radiotelephone distress frequency watch receiver preset on 2182 kHz and provided with loudspeaker silencing device (until 1.2.99).	**S92** IV/7.2				
14	Device for generating the radiotelephone alarm signal (2182 kHz) (until 1.2.99).	**S92** IV/7.3				
15	Secondary means of ship-to-shore alert from the bridge by a radio service (A1: MF DSC, A2: non-MF DSC, A3: on HF DSC, A4: HF INMARSAT). A2: may be used as a secondary means of ship-to-shore alert from the bridge by a radio service other than MF operating through the INMARSAT service or a separate item should be provided	**S92** IV/8.1.3 **S92** IV/8.1.4 **S92** IV/9.1.3.2 **S92** IV/10.1.4.2 **S92** IV/9.1.3.3.1 **S92** IV/11				
16	Emergency position-indicating radio beacon (VHF EPIRB) installed in easily accessible position, ready to be manually released and activated, capable of floating free and being automatically activated when afloat and capable of transmitting a distress alert in DSC on VHF channel 70 and capable of locating 9 Ghz signals, if ship engaged on voyages only in sea area A1. 3: may be used as a secondary means of ship-to-shore alert from bridge operating on VHF using DSC or a separate item should be provided.	**S92** IV/8.3 **S92** IV/8.1.1				
17	MF T+R operating from the bridge on 2187.5 kHz using DSC and on 2182 kHz using RTP, capable of initiating distress alerts from bridge. Incorporated MF DSC watch receiver on 2187.5 kHz (may be separated).A2 5: T+R for general radiocommunications using RTP or direct-printing telegraphy by a radio installation operating between 1605 and 4000 kHz or	**S92** IV/9.1.1 **S92** IV/9.1.2 **S92** IV/9.3.1 **S92** IV/10.1.2 **S92** IV/10.1.3				

No	Guidelines	Ref.	A1	A2	A3	A4
	between 4000 and 27500 kHz (may be separated).					
18	INMARSAT ship earth station (A1: as a secondary means of ship-to-shore alert from bridge, A2: as a T+R for general radiocommunications using RTP or directprinting telegraphy, A3: capable of transmitting and receiving distress and safety communications from bridge using direct-printing telegraphy, initiating and receiving distress priority calls, maintaining watch for shore-to-ship distress alerts, transmitting and receiving general radiocommunications using either RTP or direct-printing telegraphy.	**S92** IV/8.1.5.1 **S92** IV/9.3.2 **S92** IV/10.1.1				
19	MF/HF T+R for distress and safety communications using DSC and RTP and direct-printing telegraphy by a radio installation operating from bridge between 1605 and 27500 kHz. Incorporated MF/HF DSC watch receiver on 2187.5 kHz, 8414.5 kHz and at least one of the DSC distress and safety frequencies (4207.5 kHz, 6312 kHz, 12577 kHz or 16804.5 kHz). At any time it shall be possible to select any of the frequencies (may be separated). Incorporated T+R for general radiocommunications using RTP or direct-printing telegraphy by an MF/HF radio installation operating between 1605 and 27500 kHz (may be separated).	**S92** IV/10.2.1 **S92** IV/10.2.2 **S92** IV/10.2.4 **S92** IV/11				

SECTION N - CHEMICAL TANKERS

Guidelines: Where the text is underlined, it does not refer to the BCH Code.

No	Item	Ref.	Guidelines
1	Wheelhouse doors, windows, sidescuttles and windows in superstructure and deck-house ends facing the cargo area satisfactory	**IBC**/3.2.3 **BCH**/2.7.3	Entrances, air inlets and openings to accommodation, service and machinery spaces and control stations should not face the cargo area. They should be on the outboard side of the superstructure at least 4% L from the end bulkhead facing the cargo area (but in no case less than 3 m and more than 5 m). No doors permitted within these limits, apart from doors leading to spaces separated from accommodation and with A-60 boundaries. Bolted plates permitted. Wheelhouse openings permitted provided that fast tightening is ensured. Windows and sidescuttles permitted provided they are nonopening type and, in first tier, they have steel covers.
2	Access ladders to cpr satisfactory	**IBC**/3.3.1-4 **BCH**/2.8.1-5	Unrestricted passage to be ensured at all times from any ladder handling ensured. Guardrailings to be fitted on all ladders and platforms. Normal access ladders not vertical, with platforms at suitable intervals. Permanent arrangements should be made for hoisting an injured person with a rescue line, avoiding obstacles.
3	Removable pipe lengths and equipment for cargo separation available in cpr	**IBC**/3.5.2 **BCH**/2.21.2	Filling of ballast in cargo tanks permitted from permanent ballast pumps through spool pieces and non-return valves.
4	Sealing arrangements of all penetrations of cpr bulkheads satisfactory	**IBC**/3.3.7	Where machinery is driven by shafting through bulkheads/decks, gastight seals with efficient lubrication should be fitted in way of the bulkhead/deck.

No	Item	Ref.	Guidelines
5	Test remote operation of cpr bilge system	**IBC**/3.3.5 **BCH**/2.8.6	The bilge system serving the cpr should be operable from outside the cpr. Cpr drainage should go to one or more slop tanks provided with standard coupling shore connection.
6	Check bilge and ballast arrangements and piping	**IBC**/3.5.1, 3 **BCH**/2.21.1, 22	Ballast pumps, lines, vent lines and similar for permanent ballast tanks to be independent of cargo system. Filling permitted from machinery spaces only from tank deck level through non-return valves. Bilge pumping arrangements for spaces within the cargo area should be situated inside this area apart for void spaces, double bottom tanks, ballast tanks separated from cargo tanks by a double bulkhead.
7	Check bow/stern loading/unloading arrangements, test means of communication and remote shutdown for cargo pumps	**IBC**/3.7, 11.3.16 **BCH**: see rules for cargo area	Not permitted for products carried by type 1 ships. Piping outside cargo area at least 760 mm inboard, on open deck, clearly identified and fitted with shutoff valve, spool piece, blank flanges, spray shields and collecting trays at its connection to the cargo piping system within the cargo area. Shore connection provided with shutoff valve and blank flange. Flanges permitted only within cargo area and at shore connection. Purging and gas-safe maintaining to be possible through shutoff valve and blank flange. Entrances, air inlets and openings to accommodation, service and machinery spaces and control stations: see N1. Continuous coamings of suitable height to be fitted for keeping spills on deck away from accommodations. Escape routes not to terminate within 3 m from

No	Item	Ref.	Guidelines
			the coamings. Electric equipment within 3 m to be safe type. One foam monitor to protect shore connection and one portable applicator to protect the pipe to be provided. Means of communication between ccr and shore connection to be provided. Remote shutdown of cargo pumps to be fitted nearby the shore connection.
8	Check cargo transfer arrangements and suitability of hoses	**IBC**/3.3.6, 3.6, 5.5-7 **BCH**/2.8.7, 2.10.3-7, 2.11, 2.12, 2.23	Pumps, valves and pipelines to be identified with service and tanks served. Pump discharge pressure gauges to be provided outside the cpr. In any cpr where a pump serves more than one tank, a stop valve shall be fitted on each line. One stop valve capable of being manually operated to be fitted on each tank filling and discharge line. One stop valve to be fitted at each cargo hose connection. Remote shutdown devices available for each cargo pump. Liquid and vapour hoses to be marked with maximum working pressure (not less than 10 bar), maximum and minimum service temperature.
9	Check cargo heating/cooling system, sampling arrangements, means for measuring temperature and alarms	**IBC**/7.1 **BCH**/2.15	Valves to be provided for each tank to isolate and regulate flow manually. Means for measuring temperature to be provided according to particular requirements for each product. Types accepted: restricted (portable thermometer lowered into a gauge tube of the restricted type) or closed (remote reading thermometer with sensor installed in the tank). Toxic products: system completely separated from other systems and not entering

No	Item	Ref.	Guidelines
			machinery spaces; arrangements in cargo area for sampling.
10	Check cargo tank vent system, P/V valves, devices for preventing passage of flame, drain lines	**IBC**/8.1, 2 **BCH**/2.13, 14	Liquid head in any tank due to venting system not to exceed the test head. To this aim, high level alarms, overflow control system, spill valves, automatic closing valves together with gauging devices to be provided. Flame screens on discharge openings to be easily accessible and removable for cleaning. Vent lines fitted with drainage system. Open vents or controlled vents with pressure/vacuum valves can be provided for each tank or commonly if segregation not required, shutoff valves not accepted. Height of vent outlets not less than 4 m above deck or gangway (if within 4 m from outlets). Height can be reduced to 3 m if high velocity valves are fitted on the outlets. Vent outlets to be at least 10 m from air intakes or openings of accommodation, service and machinery spaces. Additional requirements for ships constructed on or after 1.1.94: master to be provided with maximum permissible loading and unloading rates for each tank. Height of vent outlets not less than 6 m above deck or gangway (if within 4 m from outlets). Height can be reduced to 3 m if high velocity valves are fitted on the outlets. Gas freeing accepted through ship's outlets or portable hoses with high velocity valves, at a minimum height of 2 m above deck.

No	Item	Ref.	Guidelines
11	Check gauging devices, high-level alarms, overflow control valves	**IBC**/13.1, 15.19 **BCH**/3.9-10, 4.14	Specific requirement for certain products. Alarm for power failure of systems essential for safe loading to be provided. Visual and audible high level alarm to be fitted, provided with power failure alarm and capable of being tested prior to loading. HL alarm to indicate approaching to normal full condition. Overflow control system to be fitted to come into operation for failure of normal procedure to stop tank filling, giving visual and audible alarm. A signal system for onshore personnel even dependent on operator's intervention to be provided. Use of shipboard automatic closing valves accepted upon Administration approval. Gauging system, HL alarm and overflow control system to be independent. BCH Code provides an alternative to the above: high level alarm to be fitted, capable of being tested prior to loading. Automatic overflow control system to be fitted to come into operation for failure of normal procedure to stop tank filling. When automatic closing valves are included, they shall be fail-safe type. HL alarm and overflow control system to be independent. Gauging systems can be different depending on the type of product: open (ullage opening), restricted (emitting small quantity of vapour, normally closed, avoiding spraying during opening), closed (float-type, electronic / magnetic probe, protected sight glass) Open and restricted types need arrangements for relieving tanks pressure before operating the gauge.

No	Item	Ref.	Guidelines
12	Check arrangements for sufficient gas carried/generated for normal losses compensation	**IBC**/9.1.3.1 **BCH**/2.19.3(a)	An adequate supply of inert gas or padding medium for use in filling and discharging the cargo tanks should be carried or manufactured on board unless a shore supply is available. In addition, sufficient inert gas or padding medium should be available on the ship to compensate for normal losses during transportation.
13	Check means for monitoring ullage spaces	**IBC**/9.1.3.4 **BCH**/2.19.3(d)	Means should be provided for monitoring ullage spaces containing a gas blanket to ensure that the correct atmosphere is being maintained.
14	Check arrangements for sufficient medium carried where drying agents are used on air inlets of cargo tanks	**IBC**/9.1.4 **BCH**/2.19.3(f)	Where drying agents are used as the drying medium on all air inlets to the tank, sufficient medium should be carried for the duration of the voyage, taking into consideration the diurnal temperature and the expected humidity.
15	Check electrical equipment in dangerous zones for maintenance and suitability, pipelines and independent cargo tanks electrically bonded	**IBC**/10 **BCH**/3.3-8	The use of intrinsically safe systems and circuits (ISS) in all hazardous locations is always accepted and particularly recommended for measurement, monitoring, control and communication purposes. Non-flammable cargoes: submerged cargo pumps motors and cables may be accepted provided that a shutdown device with alarm is fitted for low liquid level alarm. Electrical equipment in cpr to be non-sparking type. When cargo is heated within 15°C from its flashpoint, cpr and areas within 3 m are hazardous areas and equipment to be safe type. Flammable cargoes: in

No	Item	Ref.	Guidelines
			addition to ISS, the following equipment is accepted: cables (open decks, holds containing cargo tanks), cables in heavy gauges gastight steel pipes (void spaces), depth sounding and log devices in gastight enclosures (void spaces, holds containing cargo tanks), lighting with pressurised enclosures or flameproof type with switches in safe area (holds containing cargo tanks, cpr), flameproof general alarm indicators (cpr), safe-type certified equipment (open decks). Independent cargo tanks to be electrically bonded to the hull. All gasketed cargo pipes and hose connections to be electrically bonded. Apart from what indicated above (not underlined) BCH Code requires explosion proof equipment in spaces provided with forced ventilation and explosion proof lighting in cpr. Equipment of enclosed ventilated design may be used in non-hazardous areas.
16	Check fixed fire-fighting system for cpr	IBC/11.2 BCH/3.13.3-4	CO_2 (0,56 m^3/kg) available for 45% of the gross volume of the cpr. Halon 1301 available for 7% of the gross volume of the cpr. Pipe provided with a valve operable from the deck clearly marked with "cargo pump room" (clear instructions to be posted). Shutdown for ventilators and closing appliances operable from outside the cpr. Automatic sound alarm, safe for flammable products, provided before discharge (suitably delayed). Bottle

No	Item	Ref.	Guidelines
			storage room to be in safe and accessible position and effectively ventilated, separated from cpr and with gastight boundaries and outwards opening door. Means provided to crew for checking the content of bottles. Notice outside: system only for fire fighting and not for inerting. Water spray and high expansion foam may be accepted by the Administration for certain cargoes (specified in ICOF/COF).
17	Check deck foam system for cargo area	**IBC**/11.3 **BCH**/3.14	Only one type of foam provided, suitable for maximum number of cargoes. Additional requirements for other cargoes according to Administration. Main control station outside of cargo tank area, adjacent to accommodation, readily accessible and operable. Rate of supply not less than 2 litres per minute per square metre (ship's breadth x cargo tanks total length) or 20 litres per minute per square metre (horizontal area of the greatest cargo tank) or 10 litres per minute per square metre (area protected by largest monitor, not less than 1250 l/minute) (the greater) for at least 30 minutes. Each monitor to provide at least 50% required froth rate. A monitor and a hose connection for a froth applicator provided each side accommodation front facing cargo area. At least 4 applicators to be provided, capacity 400 l and throw 15 m. Valves provided in both the froth main and the fire main, fwd each

No	Item	Ref.	Guidelines
			monitor. Operation of froth system at nominal output to allow contemporary use of fire main with two jets of water.
18	Check arrangements for ventilation of spaces normally entered during cargohandling operations and other spaces in the cargo area, spares for mechanical ventilation fans	**IBC**/12 **BCH**/3.1-2	Cpr and other enclosed spaces containing cargo handling equipment to have mechanical ventilation. Notice outside: ventilate before entering. Accumulation of vapours to be avoided, not less than 30 changes per hour to be provided. Ventilation shall be always extraction type. Spaces for motors running cargo pumps shall have positive pressure. Exhaust ducts at least 10 m from air intakes and openings of accommodation, service and machinery spaces. Electric motors outside the ventilated space, fans to be non-sparking type. Sufficient spare parts for each fan. Protection screens of not more than 13 mm square mesh fitted in outside openings of ducts. Pump rooms and other spaces normally entered, to have mechanical ventilation controlled from outside 20 changes per hour. Spaces not normally entered: fixed ventilation / portable (8 / 16 changes per hour). Electric motors outside the ventilated space, fans to be non-sparking type.
19	Check instrumentation	**IBC**/13.2 **BCH**/3.11.1-2	Ships carrying toxic or flammable products, to have at least two portable instruments designed and calibrated for testing the specific vapours. If a fixed system is provided, at least one portable instrument is required.

No	Item	Ref.	Guidelines
20	Check equipment for personnel protection: protective clothing, safety equipment and associated breathing apparatus, emergency-escape respiratory and eye protection, medical first-aid equipment, stretchers, oxygen resuscitation equipment, antidotes for cargoes, decontamination arrangement and eyewashes, gas-detection instruments, stowage of cargo samples	**IBC**/14 **BCH**/3.16	Crew members engaged in cargo operations provided with protective equipment (large aprons, special gloves with long sleeves, suitable footwear, coveralls of chemical-resistant material, tight-fitting goggles or face shields. Protective equipment to be kept in suitable lockers easily accessible, not in accommodation unless segregated. Ships carrying toxic products shall carry at least three safety equipment (In addition to SOLAS requirements) including: one breathing apparatus, protective clothing, boots, gloves, tight-fitting goggles, fireproof lifeline with resistant belt, ex-proof lamp. Each breathing apparatus shall have fully charged spare air bottles with 6000 l free air capacity or one set of fully charged air bottles and a special air compressor with charging manifold. A cpr of a ship carrying toxic products shall have a low pressure line to be used with breathing apparatus without using its air bottles for 1 hour or equivalent spare bottled air. Safety equipment to be kept in suitable, clearly marked, easily accessible places (one near cpr entrance). A stretcher for hoisting an injured person should be placed in a readily accessible position. Certain cargoes need filter-type respiratory protection, emergency escape respiratory protection for at least 15'. Medical first-aid equipment with oxygen resuscitation and antidotes for cargoes to be provided. Suitably marked decontamination showers and

No	Item	Ref.	Guidelines
			an eyewash to be available on deck, operable in all ambient conditions.

SECTION O - GAS CARRIERS

Guidelines: Where the text is underlined, it does not refer to the GC Code; where the text is in italics it does not refer to the GC ex Code.

No	Item	Ref.	Guidelines
1	Wheelhouse doors, windows, sidescuttles and windows in superstructure and deck-house ends facing the cargo area satisfactory	**IGC**/3.2.4-6 **GC**/3.2.4-6 **GC ex**/3.2.2-4	Entrances, air inlets and openings to accommodation, service and machinery spaces and control stations should not face the cargo area. They should be on the outboard side of the superstructure at least 4% L from the end bulkhead facing the cargo area (but in no case less than 3 m and more than 5 m). No doors permitted within these limits, unless non-opening type. Wheelhouse openings permitted provided that fast tightening is ensured. Windows and sidescuttles permitted provided they are non-opening type. All air intakes and openings to be fitted with closing devices (operable from the inside where toxic gases are carried).
2	Check cargo pump rooms and cargo compressor rooms	**IGC**/3.3 **GC**/3.3 **GC ex**/3.3	Where pumps and compressors are driven by shafting passing through a bulkhead / deck, gastight seals with efficient lubrication or equivalent should be fitted in way of the bulkhead / deck. Safe and unrestricted access to be ensured to personnel wearing protective clothing and breathing apparatus. Removal of unconscious / injured personnel to be allowed. All

No	Item	Ref.	Guidelines
			cargo handling valves to be accessible. Suitable means of drainage to be provided.
3	Check manually operated and automatic emergency shutdown of cargo pumps and compressors	**IGC**/5.6.1.3	Shutdown of pumps / compressors to be automatic where the emergency shutdown valves in O-12 are shut automatically by fusible element or by fail-safe closure.
4	Check cargo control room (ccr)	**IGC**/3.4 **GC**/3.4 **GC ex**/3.4	To be a gas-safe space if within accommodation, service spaces and control stations. Access to accommodation, service spaces or control stations admitted granted that entrances to ccr comply with O-1, otherwise access forbidden and boundaries to be A-60. If gas-safe, instrumentation to be indirect reading system, preventing gas escape. If not gas-safe and ship carries flammable products, sources of ignition to be excluded and electrical installations to be safe-type.
5	Check gas-detection equipment	**IGC**/9.1.2-2.2.2, 13 **GC**/9.1.2-2.2(b), 13 **GC ex**/9.1.2-2.2 (b), 13	A sufficient number of sampling point for checking operation of cargo tanks and piping gas-freeing and purging to be provided. Gas sampling connections to be valved on main deck. Detection of leakages from tanks to be also available. Audible and visual alarms should be located on navigating bridge, remote control position for cargo valves and at the gas detector readout position. Where gas detection system is located in a gas-safe space pipes should have shutoff valves and exhaust gas from detector should be discharged

No	Item	Ref.	Guidelines
			in safe location. Suitable equipment and span gas for periodical calibration to be available. A permanent gas detection system to be provided for: cpr, cargo compressor room, motor room, ccr unless gas-safe, other enclosed spaces in cargo area, ventilation hoods and ducts, air-locks. Sampling frequency for each space not to exceed 30 min (hoods and ducts: continuous). For some cargoes, portable detectors in lieu of fixed system are accepted. Alarm should come out for 30% of lower flammable limit.s
6	Check arrangements for air-locks	**IGC**/3.6 **GC**/3.6 **GC ex**/3.6	Permitted between gas-dangerous zone on open deck and gas-safe space. Consisting of two steel, gastight doors spaced at least 1,5 m but not more than 2,5 m. Doors to be self-closing and without hold back hook. Audible and visual alarm both sides indicating more than one door open. Air-lock to be mechanically ventilated from a gas-safe space and to be overpressure to the gas-dangerous zone. To be monitored for cargo vapour. Doors sill not less than 300 mm. In ships carrying flammable products, electrical equipment not safe-type in spaces protected by air-lock to be de-energised upon loss of overpressure (GC ex: only motors in cargo motor rooms). Non safetype equipment for manoeuvring, anchoring, mooring, fire-fighting not admitted in spaces protected by air-lock.

No	Item	Ref.	Guidelines
7	Check bilge, ballast oil fuel arrangements	**IGC**/3.7 **GC**/3.7 **GC ex**/3.7	Hold spaces or insulation spaces, where fitted, to be provided with drainage system not connected with the machinery spaces. Detection system of leakages to be provided. These systems are not required for internal insulation tanks and interbarrier spaces filled by insulation material. Drainage system for interbarrier spaces should provide return to cargo tanks. Ballast spaces, fuel oil tanks and gas-safe spaces may be connected to the machinery spaces. Pump vents should not be open to machinery spaces.
8	Check bow/stern loading/unloading arrangements, test means of communication between ccr and shore	**IGC**/3.8, 11.3.1.3, 11.4.7 **GC**/3.8, 11.4.7 **GC ex**/3.8	Not permitted for products carried by type 1G ships. Piping outside cargo area at least 760 mm inboard, on open deck, clearly identified and fitted with shutoff valve, spool piece, blank flanges at its connection to the cargo piping system within the cargo area. Flanges permitted only within cargo area and at shore connection. Purging and gas-safe maintaining to be possible through shutoff valve and blank flange. Entrances, air inlets and openings to accommodation, service and machinery spaces and control stations: see O-1. Deck openings and air inlets within 10 m from the cargo shore connection to be kept closed while using the bow/stern arrangement. Electric equipment within 3 m to be safe type. One additional dry chemical powder unit with one monitor and one hand hose line to protect the pipe and shore

No	Item	Ref.	Guidelines
			connection to be provided. Additionally, a water spray system to be provided where flammable and/or toxic cargoes are carried. Means of communication between ccr and shore connection to be provided. GC ex: to the satisfaction of the Administration.
9	Check cargo and process piping, expansion arrangements, insulation from the hull structure, pressure-relief and drainage arrangements	**IGC**/5.2 **GC**/5.2 **GC ex**/5.2	Offsets, loops, bends and mechanical expansion joints (bellows, slip joints and ball joints) to be provided to protect the piping from excessive stresses due to thermal and structure movements. Low temperature piping to be isolated from hull structure. Where dismantling is frequent or leakage is predictable (shore connections, pump seals), protection for the hull beneath to be provided. Pipelines or components that may be isolated in a liquid full condition, to be provided with relief valves connected with cargo tanks or vent system (if liquid detection system provided).
10	Check cargo tank/interbarrier space pressure and relief valves, safety systems and alarms	**IGC**/5.6 **GC**/5.3 **GC ex**/5.3	Where MARVS $\leq$ 0,7 bar, all liquid and vapour connections except safety relief valves and liquid level gauging devices, should have shutoff valves close to the tank. The valves may be remotely controlled but should be capable of local manual operation. Additionally, one remotely controlled emergency shutdown valve between ship and shore to be provided. Where MARVS 〉 0,7 bar, the same connections should be equipped with a manually operated stop valve and a

No	Item	Ref.	Guidelines
			remotely controlled emergency shutdown valve located close to the tank. One remotely operated emergency shutdown valve should be provided at each cargo hose connection in use. Emergency shutdowns valves to be operable by single controls in at least two remote locations (one being ccr or the remote control location for cargo valves, if provided). Fusible elements for automatic shutdown for fire to be provided. Emergency shutdown valves to be fail-safe (loss of power) and capable of closing in 30 s.
11	Check liquid/vapour hoses for suitability	**IGC**/5.7 **GC**/5.4 **GC ex**/5.4	To be suitable for cargo and temperature. To be permanently marked with maximum working pressure (not less than 10 bar), maximum and minimum temperature.
12	Check cargo pressure/temperature control	**IGC**/7 **GC**/7 **GC ex**/7	Maintenance below the MARVS provided by: mechanical refrigeration, boil-off cargo vapours used as shipboard use fuel / waste heat or other equivalent systems approved by the Administration. A refrigeration system should consist at least of a unit and of an equivalent stand-by unit arranged in way that evaporated cargo is compressed, condensed and returned to cargo tanks or cooled / condensed without being compressed or compressed / condensed and returned to cargo tanks.

No	Item	Ref.	Guidelines
13	Check cargo, bunker, ballast and vent piping, vent masts and protective screens	**IGC**/8 **GC**/8 **GC ex**/8	Each cargo tank 〉 20 m^3 to have two pressure relief valves (prv) (one accepted for lesser volume). Interbarriers: according to Administration. Prv are to be set and sealed by the Administration and a record must be kept on board. Where two settings are admitted, it may be accomplished by installing two set and sealed valves capable of being isolated when not in use or providing means not requiring testing after changing setting. Change under Master's responsibility and recorded in the log-book. Stop valves before prv admitted provided that not more than one out of service, automatic indication of the excluded prv and nominal relieving capacity performed by the remaining prv are granted. Each prv to be connected to a venting system with exits height ≥ B/3 or 6 m (the greater) above the weather deck. Prv exits ≥ B or 25 m (the greater) from air intakes or openings of accommodations, service and control stations (other exits ≥ 10 m). Drainage to be provided. Vent outlets with suitable screens. Depending on cargo tanks scantling, a vacuum protection system may be provided (two independent pressure switches to sequentially alarm and stop all cargo suctions, vacuum relief valves suitably sized and set, or other equivalent arrangements). The system should be capable of being tested.-

No	Item	Ref.	Guidelines
14	Check arrangements for sufficient inert gas carried/generated for normal losses compensation, spaces monitoring	**IGC**/9.4-5 **GC**/9.4-5 **GC ex**/9.3-4	Inert gas may be provided from the shore or from the ship. Inert gas for fire-fighting to be separated from that used for environmental control. Backflow of cargo vapour to igs to be prevented. Each inerted space capable of being isolated and pressure control available. O2 continuous reading fitted on the igs supply, with an alarm for O2 exceeding 5%. Igs room and piping not inside accommodations, service spaces and control stations but may be in machinery spaces (provided that two non-return valves are fitted). Inert gas to be sufficient for 30 days of normal consumption.
15	Check air drying system, interbarrier and hold space purging IGS	**IGC**/9.1-3 **GC**/9.1-3 **GC ex**/9.1-2	A system for gas-freeing and purging to be available for cargo tanks and piping. Where permitted by the cargo nature, dry air may be used in lieu of inert gas (provided that a reserve for inerting the largest space is available).
16	Check electrical equipment in dangerous zones for maintenance and suitability, pipelines and independent cargo tanks electrically bonded	**IGC**/5.2.1.4, 10 **GC**/5.2.3, 10 **GC ex**/5.2.3, 10	Electrical bonding to the hull of insulated piping and tanks and between gasketed joints to be provided. Intrinsically safe electrical equipment and wiring may be fitted in all gas-dangerous spaces. Submerged cargo pumps and wiring may be fitted in cargo tanks. In case of low liquid level, the motors should shut down automatically, giving an alarm in ccr. Cargo pump motors capable of being isolated from electrical supply during gas-freeing. Supply cables for submerged cargo pumps may be installed in holds for tanks with interbarrier. Where interbarrier is not

No	Item	Ref.	Guidelines
			needed, the following may be installed: cables, flameproof type lighting with switches in gas-safe area, electrical depth sounding or log devices in gastight enclosures. Equipment of cargo pumps and compressors rooms: flameproof type lighting with switches in gas-safe area, electric motors for cargo pumps and compressors to be separated by a gastight bulkhead / deck (gastight seals with efficient lubrication or equivalent should be fitted in way of the bulkhead / deck), general alarm audible indicators in flameproof enclosures. Open decks: within 3 m from cargo tanks outlets, cargo pipe flanges, cpr openings, compressor room openings, etc. and for 2,4 m height from deck only cables and certified safe type equipment are accepted. Enclosed spaces containing cargo arrangements: cables and flameproof type lighting. Spaces protected by air-locks: certified safe type equipment unless capable of being de-energised (see O-6).
17	Test the remote means of starting of one fire pump	**IGC**/11.2.5 **GC**/11.2.5 **GC ex**/11.2.5	Where the engine room is unattended, at least one fire pump should be started and connected to the fire main by remote control from bridge or other control station outside the cargo area.
18	Check fixed fire-fighting system for gasdangerous spaces, means of operation clearly marked	**IGC**/3.3.1.1, 11.5 **GC**/3.3.1(a), 11.5 **GC ex**/11.5.1, 11.5	Enclosed spaces normally entered where flammable liquid or vapour leakage may occur, such as cargo pumps and compressors rooms, to be protected by a fixed installation (see J-3 and J-4).

No	Item	Ref.	Guidelines
19	Test water-spray system for cooling, fire protection, crew protection, means of operation clearly marked	**IGC**/11.3 **GC**/11.3 **GC ex**/11.3	Compulsory for ships carrying flammable and/or toxic products. It should cover: exposed cargo tanks / deck storage vessels and domes, cargo liquid and vapours manifolds, boundaries of superstructures normally manned, cargo compressor and cargo pumps rooms all facing the cargo area. Minimum capacity: 10 l/m^2 per minute (horizontal surfaces), 4 l/m^2 per minute (vertical surfaces). Stop valves to be provided for isolating damaged sections or the system could be divided into sections, independently operated. A dedicated water spray pump could be provided or the fire pumps may be used provided that their capacity is increased by the amount needed for the water spray system.
20	Check deck dry chemical powder system, means of operation clearly marked	**IGC**/11.4 **GC**/11.4 **GC ex**/11.4	Compulsory for ships carrying flammable products. Capable of delivering powder from two hand hose lines or monitors to any part of deck cargo area. The system should be activated by nitrogen or similar. The system should consist of at least two independent selfcontained units (one for cargo capacity 〈 1000 m^3). A monitor to be provided for manifolds, capable of being remotely actuated (not required if satisfactory coverage provided from a single position. All hoses and monitors provided with local actuator. One hose line or monitor to be located aft the cargo area. Monitor capacity not less than 10 kg/s (3,5 for hand hose lines). Each unit to have sufficient powder for all the connected monitors and hose lines for 45 s.

No	Item	Ref.	Guidelines
21	Check equipment for personnel protection: two complete sets of safety equipment for entering gas-filled spaces, supply of compressed air and special air compressor (if provided), emergencyescape respiratory and eye protection, medical first-aid equipment, stretchers, oxygen resuscitation equipment, decontamination arrangement and eyewashes, arrangements to protect personnel against cargo release	**IGC**/11.6, 14 **GC**/11.6, 14 **GC ex**/11.6, 14	All ships carrying flammable products should carry fireman's outfits (see J-1.14/15) according to its cargo capacity as follows: 2 (〈 2000 m^3), 4 (between 2000 and 5000 m^3), 5 (〉 5000 m^3). GC ex: 3 outfits and sets of protective clothing (〈 25000 m^3), 5 outfits and sets ($\geq$ 25000 m^3). All breathing apparatus to be self-contained type with 1200 l free air minimum capacity. In addition to be provided at least two safety equipment (1200 l In addition to be provided at least two safety equipment (1200 l breathing apparatus, protective clothing, boots, gloves, tight-fitting goggles, rescue line with belt, ex-proof lamp) with adequate supply of compressed air (one set of charged bottles for each BA and recharge compressor with manifold and spare air bottles or 6000 l charged air bottles for each BA). As an alternative to air supply, to be provided a low pressure air line in gas-dangerous spaces for connection to BA and recharge compressor. Suitable protective equipment for personnel engaged in loading operation to be provided. Safety and protective equipment to be kept in suitable, clearly marked lockers located in readily accessible places. Compressed air equipment to be inspected once a month by a responsible officer and recorded in log-book, and tested by an expert once a year. A hoistable stretcher to be kept in readily accessible location. Medical

No	Item	Ref.	Guidelines
			first-aid equipment including oxygen resuscitation equipment to be provided. For some products, the following to be provided: respiratory and eye protection for each person on board (filter type respiratory or 15 min SCBA plus two on navigating bridge), suitably marked decontamination showers and an eyewash on deck, operable in all ambient conditions, two (GC ex: three) additional safety equipment with three spare charged air bottles (cargo capacity ≥ 2000 m^3).
22	Check arrangements for ventilation of spaces normally entered during cargohandling operations and other spaces in the cargo area, spares for mechanical ventilation fans	**IGC**/12 **GC**/12 **GC ex**/12	Electric motors, cargo pumps and compressors rooms to be fitted with a mechanical ventilation system operable from the outside. Spaces to be ventilated before entering (warning notice to be placed outside). Minimum capacity: 30 (GC ex: 20) air changes per hour (8 for gas-safe ccr). Gas-safe spaces to have positive pressure type system, gas-dangerous spaces: negative. Gas-dangerous spaces outlets to be at least 10 m from accommodations, service spaces and control stations openings. Electric motors for fans to be outside the ducts if flammable products are carried, fans to be non-sparking type. Spare parts for each type of fan to be available. Protection screens of not more than 13 mm square mesh to be fitted on outlets. Spaces not normally entered should have at least portable ventilation system, non-sparking type.

No	Item	Ref.	Guidelines
23	Check liquid level indicators, pressure gauges, high/low pressure alarms, overflow control valves, temperature indicators	**IGC**/13 **GC**/13 **GC ex**/13	Each cargo tank to have level, pressure and temperature indicators to be fitted near remote controls of cargo valves, if provided. Instruments to be tested and recalibrated at regular intervals according to approved procedures. Level indicators to be of the following types (according to cargo nature): indirect, closed (penetrating the tank or not), restricted. Each cargo tank shall have a high liquid level alarm independent from indicators. Another independent sensor shall operate a shutoff valve to avoid overpressure and tank becoming liquid full (see O-12). HL alarm not required for pressure tanks 〈 200 m^3. Level alarms electrical circuits capable of being tested before loading. Each cargo tank to have a high pressure alarm and, where vacuum protection is required, a low pressure alarm in navigating bridge. Tanks fitted with multiple set relief valves to have a HP alarm for each set. Each cargo pump and liquid / vapour manifold to have a pressure gauge. Each cargo tank to have two temperature indicators, one in the lower part one in the upper.
24	Check two sets of portable gas-detection equipment, one suitable instrument for measuring oxygen level	**IGC**/13.6.13-14 **GC**/13.6.13-14 **GC ex**/13.6.12-13	Every ship should have two portable gas detectors and one portable oxygen-meter.
25	Check arrangements for use of cargo as fuel (if provided): gas supply cut-off for exhaust ventilation failure, master gas	**IGC**/16 **GC**/16 **GC ex**/16	Use of cargo as fuel is permitted only for methane (LNG). Gas fuel lines not to pass inside accommodations, service spaces or control stations

No	Item	Ref.	Guidelines
	fuel valve remote closure from machinery space		unless double wall pipe (space between two pipes filled with inert gas at a pressure greater than that of fuel gas and alarm for loss of pressure) or pipe inside ventilated duct (pressure inside duct lower than atmospheric, 30 changes per hour, external fans, outlets far from ignition sources, gas and air pressure detection with alarm and shutdown of gas fuel). Each gas utilisation unit to be provided with three automatic valves: two in series in the gas fuel pipe and one on a pipe venting the pipe between the two a.m. valves. Failure of forced draft, loss of flame, abnormal fuel gas pressure, failure of the valve control system should cause the two valves to close and the third to open. A master valve on the fuel gas supply operable from inside the machinery space to be fitted outside it. This valve should close for detection of gas, duct ventilation loss, double wall pipe pressure loss. Pipes in machinery space fitted with inerting and gas-freeing systems.

Annex 1
Supplies from the emergency source

No	Item	Required time			
		S 92	S 81	S 74	S 60
1	Emergency lighting at every muster and embarkation station and over the sides	3h	3h	6h	6h
2	Emergency lighting in service and accommodation alleyways, stairways and exits and (S 81, S92) personnel lift cars and personnel lift trunks	18h	18h	6h	6h
3	Emergency lighting in machinery spaces and main generating stations	18h	18h	6h	6h
4	Emergency lighting in all control stations, control rooms main and emergency switchboards (S 74: only navigating bridge)	18h	18h	6h	6h
5	Emergency lighting at all stowage positions for fireman's outfits	18h	18h	-	-
6	Emergency lighting at the steering gear	18h	18h	-	-
7	Emergency lighting at the fire emergency pump	18h	18h	-	-
8	Navigation lights	18h	18h	6h	6h
9	VHF radio installation (IV/7.1.1, 7.1.2) (1.2.95)	18h	18h	-	-
10	MF radio installation (IV/9.1.1, 9.1.2, 10.1.2, 10.1.3)	18h	-	-	-
11	Ship earth station	18h	-	-	-
12	MF/HF radio installation (IV/10.2.1, 10.2.2, 11.1)	18h	-	-	-
13	All internal emergency communication (a)	18h	18h	-	-
14	The shipborne navigational equipment (a)	18h	18h	-	-
15	Fire detection and alarm system (a)	18h	18h	-	-
16	Daylight signalling lamp (a)	18h	18h	6h	6h
17	Ship's whistle (a)	18h	18h	-	-
18	Manually operated call points (a)	18h	18h	-	-
19	Emergency internal signals (a) (S 74: only general alarm)	18h	18h	6h	6h
20	Emergency steering gear	(b)	(b)	-	-

(a) Unless they have an independent supply for 18 hours.

(b) Ships $\geq$ 10000 GT: 30', other ships: 10'.

Note S 74, S 60: requirements valid for ships $\geq$ 5000 GT. Other ships need only item 1 for 3h.

Annex 2
Servicing of fire bottles/extinguishers

Permanently installed fire-fighting systems

<table>
<tr><th>Type of system</th><th>Charge check</th><th>Hydrostatic test</th><th>Test pressure</th></tr>
<tr><td>CO_2</td><td rowspan="3">content check at 2-year intervals</td><td rowspan="3">10 % of the bottles at 10-year intervals</td><td rowspan="2">25 N/mm^2</td></tr>
<tr><td>NITROGEN</td></tr>
<tr><td>HALON</td><td>1,5 times the w.p.</td></tr>
<tr><td>FOAM</td><td>analysis at 3-year intervals (twice during the 5-year period)</td><td>–</td><td>–</td></tr>
<tr><td>POWDER</td><td>smoothness test each year; quantity check at 5-year intervals</td><td>–</td><td>–</td></tr>
</table>

Portable fire extinguishers

<table>
<tr><th>Type of extinguisher</th><th>Charge check</th><th>Hydrostatic test</th><th>Test pressure</th></tr>
<tr><td>Water and foam</td><td>replacement of charge each year</td><td rowspan="2">at each special survey (5-year intervals)</td><td rowspan="2">1,5 times the w.p. (2 N/mm^2 if the w.p. is unknown)</td></tr>
<tr><td>Powder with shell not kept under pressure</td><td>check each year replacement of charge at 5-year intervals</td></tr>
<tr><td>CO_2</td><td rowspan="2">at 2-year intervals</td><td rowspan="2">extinguishers sent ashore for refilling when found empty or low charged (less than 90% for CO_2, 95% for halon and with low pressure for powder ones)</td><td>25 N/mm^2</td></tr>
<tr><td>Halon
Powder with shell kept under pressure</td><td>1,5 times the w.p. (2 N/mm^2 if the w.p. is unknown)</td></tr>
<tr><td>Small compressed air or gas bottles used in fire extinguishers not permanently kept under pressure.</td><td>internal inspection at 5-year intervals</td><td>at 5-year intervals where internal inspection is not possible</td><td>2 times the w.p. 25 N/mm^2 if of a CO_2 type with safety devices 35 N/mm^2 if of a CO_2 type without safety devices</td></tr>
</table>

Annex 3
Lifeboat fittings

No	Item	Required no.		
		S 92	S 74	S 60
1	Automatic draining valve with cap (S 92) or plug attached with lanyard	1	2	2
2	Rudder and tiller	1	1	1
3	Buoyant (S 92) lifeline around the lifeboat	1	1	1
4	Handholds underside the hull	2	2	2
5	Watertight lockers for storage of small items	suff	suff	suff
6	Means for storage of collected rainwater	suff	-	-
7	Release mechanism for hooks	1	-	-
8	Release device for forward painter	1	-	-
9	Skates and fenders (not free-fall)	suff	-	-
10	Externally manually controlled lamp on top of lifeboat	1	-	-
11	Electric lamp inside lifeboat	1	-	-
12	Permanently attached foldable canopy (partially enclosed lifeboats)	1	-	-
13	Self-contained air support system with pressure indicators (toxic cargoes)	1	-	-
14	Water spray system (inflammable cargoes)	1	-	-

Annex 4
Lifeboat inventory

No	Item	Required no.		
		S 92	S 74	S 60
1	Boarding ladder	1	1	2
2	Buoyant oars, thole pins and crutches (not free-fall lifeboats)	suff	(a)	(a)
3	Boat hooks	2	-	-
4	Buoyant (S 92) bailer	1	1	1
5	Buckets	2	2	2
6	Survival manual	1	-	-
7	Efficient compass in binnacle provided with means of illumination	1	1	1
8	Sea-anchor with (S 92) shock-resistant hawser and tripping line	1	1	1
9	Painters of 15 m/twice distance stowed lifeboat/waterline (not free-fall lifeboats)	2	2	2
10	Hatchets	2	2	2
11	Fresh water in watertight receptacles (litres per person)	3	3	3
12	Rustproof dipper with lanyard	1	-	-
13	Rustproof graduated drinking vessel	1	-	-
14	Food ration (b) in airtight packaging (ration per person)	1	1	2
15	Rocket parachute flares	4	4	4
16	Hand flares	6	6	6
17	Buoyant smoke signals	2	2	2
18	Waterproof Morse electric torch with spare batteries and bulb	1	1	1
19	Daylight signalling mirror with instructions	1	1	1
20	Copy of life-saving signals on (S 92) waterproof card/in waterproof container	1	1	1
21	Whistle	1	1	1
22	First-aid outfit in waterproof case	1	1	1

No	Item	Required no.		
		S 92	S 74	S 60
23	Anti-seasickness medicine (doses per person)	6	-	-
24	Seasickness bag (per person)	1	-	-
25	Jack-knife with lanyard and (not S 92) with tin-opener	1	1	2
26	Tin-openers	3	-	-
27	Buoyant rescue quoits with 30 m buoyant line	2	-	-
28	Manual pump	1	1	1
29	Set of fishing tackle	1	1	1
30	Tools for minor adjustment to the engine and its accessories	suff	-	-
31	Portable fire extinguisher (S 74: only motor lifeboats)	1	1	2
32	Searchlight	1	-	-
33	Radar reflector or radar transponder stowed in lifeboat	1	-	-
34	Thermal protective aids (percentage of persons) but not less than 2	10	-	-
35	Efficient 15 m painters suitable for towing (free-fall lifeboats)	2	-	-
36	Lamp with oil for 12 hours	-	1	1
37	Mast with galvanised wire stays and orange sails	-	1	1
38	Vessel with 4,5 l oil attachable to the sea anchor	-	1	1
39	Buoyant heaving lines	-	2	2
40	Cover of a highly visible colour for protection against exposure	-	1	1

(a) A single banked complement of buoyant oars, two spare buoyant oars and a buoyant steering oar; one set and a half of thole pins or crutches attached to the lifeboat by lanyard or chain, a boat hook.

(b) Ration to be 10000 kJ (S 92), to the satisfaction of the Administration (S 74, S 60), 1 kg (S 48).

suff =sufficient

Annex 5
Rescue boat fittings and inventory

No	Item	Required no.(S 92)
1	Automatic draining valve with cap or plug attached with lanyard	1
2	Rudder and tiller	1
3	Buoyant lifeline around the rescue boat	1
4	Handholds underside the hull	2
5	Release mechanism for hooks	1
6	Release device for forward painter	1
7	Watertight lockers for storage of small items	1
8	Arrangement for towing liferafts sufficient	sufficient
9	Buoyant oars or paddles to make headway in calm seas sufficient	sufficient
10	Thole pins and crutches attached with lanyard or equivalent arrangements for each oar	for each oar
11	Buoyant bailer	1
12	Efficient compass in binnacle provided with means of illumination	1
13	Sea-anchor and tripping line with hawser of adequate strength (10 m length)	1
14	Painter of sufficient length and strength attached to the release device	1
15	Buoyant line for towing liferafts (50 m length)	1
16	Waterproof Morse electric torch with spare batteries and bulb	1
17	Whistle	1
18	First-aid outfit in waterproof case	1
19	Buoyant rescue quoits with 30 m buoyant line	2
20	Searchlight	1
21	Efficient radar reflector	1
22	Thermal protective aids	10% persons / 2 (the greater)
23	Boat-hook (rigid rescue boats)	1
24	Bucket (rigid rescue boats)	1
25	Knife or hatchet (rigid rescue boats)	1
26	Buoyant safety knife (inflated rescue boats)	1
27	Sponges (inflated rescue boats)	2
28	Efficient manually operated bellows or pump (inflated rescue boats)	1
29	Punctures repair kit in suitable container (inflated rescue boats)	1
30	Safety boat-hook (inflated rescue boats)	1

Annex 6
Rigid liferafts fittings and inventory

No	Item	Required no.		
		S 92	S 74	S 60
1	Boarding ladder (S 74, S 60: at each opening)	1	1	1
2	Efficient painter	1	1	1
3	Manually controlled lamp powered by cell on top of canopy	1	-	-
4	Manually controlled lamp inside the liferaft	1	-	-
5	Lifeline around the liferaft and (S 74, S 60) inside	1	1	1
6	Painter around the liferaft	1	-	-
7	Buoyant rescue quoit with 30 m buoyant line	1	1	1
8	Knives of the non-folding type with buoyant handle and lanyard	(a)	(a)	(a)
9	Buoyant (S 92) bailers	(a)	(a)	(a)
10	Sponges	2	2	2
11	Sea-anchor with (S 92) shock-resistant hawser and tripping line	2	2	2
12	Buoyant (S 92) paddles	2	2	2
13	Tin-openers	3	3	3
14	First-aid outfit in waterproof case	1	1	1
15	Whistle	1	1	1
16	Rocket parachute flares	4	2	2
17	Hand flares	6	6	6
18	Buoyant smoke signals	2	-	-
19	Waterproof Morse electric torch with spare batteries and bulb	1	1	1
20	Radar reflector or radar transponder stowed in liferaft	1	-	-
21	Daylight signalling mirror with instructions	1	1	1
22	Copy of life-saving signals on (S 92) waterproof card/in waterproof container	1	1	1
23	Set of fishing tackle	1	1	1

No	Item	Required no.		
		S 92	S 74	S 60
24	Food ration (b) in airtight packaging (ration per person)	1	1	1
25	Fresh water in watertight receptacles (litres per person)	1.5	1.5	1.5
26	Rustproof graduated drinking vessel	1	1	1
27	Anti-seasickness medicine (doses per person)	6	6	6
28	Survival manual	1	1	1
29	Instructions for immediate action	1	-	-
30	Thermal protective aids (percentage of persons) but not less than 2	10	-	-
31	Canopy	1	1	1
32	Buoyant battery powered light with lanyard	-	1	1

(a) Liferafts permitted to accommodate 13 persons or more: 2, others: 1.

(b) Ration to be 10000 kJ (S 92), to the satisfaction of the Administration (S 74, S 60).

REFERENCES

1. Yalcin C. PSCO Training Tex Book 2004, Turkish Prime Ministry Undersecretaries of Maritime Affairs
2. IMO RESOLUTION A.787(19), PROCEDURES FOR PORT STATE CONTROL
3. DNV, Guide On Port State Control
4. NIPPON KAIJI KYOKAI, Maintenance Checklist for the Master
5. Turkish Lloyd, PSC Procedures and Implementations
6. Korean Register, Port State Control information booklet
7. Germanischer Llyod, PSC information Manual
8. PSC annual reports, http://www.parismou.org/
9. PSC annual reports, http://www.bsmou.org/
10. PSC annual reports, http://www.tokyo-mou.org/
11. PSC annual reports, http://www.amsa.gov.au/
12. USCG, http://www.uscg.mil/hq/g-m/pscweb/Index.htm

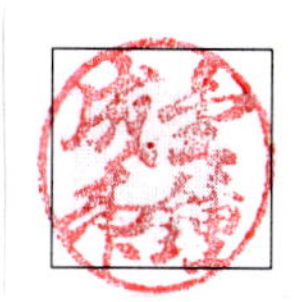

Port State Control Practice

2009년 2월 25일　1판 1쇄 발행

저　자　Volkan Aydogdu 외
발행인　남　승　우

발행처　도서출판 상尙 학學 당堂

서울특별시 동작구 사당3동 220-43
TEL : 02) 595-1692~4
FAX : 02) 595-1394
E-mail : namdoo6@unitel.co.kr

신고 : 2-155호 (1968. 11. 29)

정가 16,000원

ISBN　978-89-93383-36-2　93500